D1740195

Clifford™

Resources for the Early Years

First Letters
and Numbers

Acknowledgements

First published in the UK by Scholastic Ltd 2005

Copyright © 2002 Scholastic Entertainment Inc. All rights reserved

Based on the CLIFFORD THE BIG RED DOG book series © Norman Bridwell
published by Scholastic Inc.

Scholastic and associated logos are trademarks and/or

registered trademarks of Scholastic Inc.

CLIFFORD and associated logos are trademarks and/or

registered trademarks of Norman Bridwell.

All rights reserved.

Additional text, copyright © Scholastic Ltd, 2005

Every effort has been made to trace copyright holders for the works reproduced
in this book, and the publishers apologise for any inadvertent omissions

Published by Scholastic Ltd

Villiers House

Clarendon Avenue

Leamington Spa

Warwickshire CV32 5PR

www.scholastic.co.uk

Printed by Bell & Bain Ltd, Glasgow

1 2 3 4 5 6 7 8 9 0   5 6 7 8 9 0 1 2 3 4

British Library Cataloguing-in-Publication Data

A catalogue record for this book is available from the British Library.

ISBN 0-439-96492-x

ISBN 978-0439-96492-0

All rights reserved. This book is sold subject to the condition that it shall not, by way
of trade or otherwise, be lent, hired out or otherwise circulated without the publisher's
prior consent in any form of binding or cover other than that in which it is published
and without a similar condition, including this condition, being imposed upon the
subsequent purchaser.

No part of this publication may be reproduced, stored in a retrieval system, or
transmitted, in any form or by any means, electronic, mechanical, photocopying,
recording or otherwise, without the prior permission of the publisher. This book
remains copyright, although permission is granted to copy those pages marked
'photocopiable' for classroom distribution and use only in the school which
has purchased the book or by the teacher who has purchased the book and in
accordance with the CLA licensing agreement. Photocopying permission is given for
purchasers only and not for borrowers of books from any lending service.

Author

Dina Anastasio
Based on the Scholastic Inc book series
Clifford The Big Red Dog by Norman Bridwell

Editor

Sally Gray

Contents

■ SCHOLASTIC
PHOTOCOPIABLE

Motivate the children with these lively activity sheets that centre on the popular and well-loved characters of Clifford and his friends. The 'First letter' activities run in order from A to Z with extra activities for reinforcement. The 'First numbers' activities run from 1 to 20 with revision activities and further ideas. The notes below provide some suggestions for ways to introduce each page and extend the learning potential of the activities.

Circle the pictures (p8)

Early Learning Goal: Link sounds to letters, naming and sounding the letters of the alphabet.

What to do: Read all the words on the sheet together. Draw attention to the difference in the two types of 'a' sounds (**a**corn and **a**nt). Can the children think of other examples?

Bat and ball (p9)

Early Learning Goal: Hear and say initial and final sounds in words, and short vowel sounds within words.

What to do: Put some 'b' objects in a bag. Can the children guess what's inside – I am round and begin with 'b'. You can throw me. What am I? Play the game using the words on the sheet.

Clifford and Cleo (p10)

Early Learning Goal: Read a range of familiar and common words and simple sentences independently.

What to do: Cover each picture with a small Post-it note. Can the children point to the letter 'c'? Which word is the odd one out on each line? Help the children to sound out and read the words.

Join the dots (p11)

Early Learning Goal: Use a pencil and hold it effectively to form recognisable letters, most of which are correctly formed.

What to do: Provide younger children with pencil grips. Check that all the children are holding the pencil correctly. Encourage them to slowly join the dots. Practise writing 'd' with fingers in the air.

The eagle's eggs (p12)

Early Learning Goal: Link sounds to letters, naming and sounding the letters of the alphabet.

What to do: Discuss with older children how sounds (or phonemes) are represented by letters and how the same phoneme can be spelled in many ways. On this sheet the phonemes /ae/ (as in eight); /ee/ (as in eagle) and /e/ (as in eggs) are all used.

Fishing lines (p13)

Early Learning Goal: Use a pencil and hold it effectively to form recognisable letters, most of which are correctly formed.

What to do: Draw some straight, curvy, wiggly and zig-zag lines on a wipe-off board for the children to copy underneath. Encourage them to draw straight lines between the matching fish on the sheet.

Behind the gate (p14)

Early Learning Goal: Use a pencil and hold it effectively to form recognisable letters, most of which are correctly formed.

What to do: Help the children to trace the curvy paths with their fingers and then with a pencil. Trace the letter 'g' in the air, describing its formation before asking the children to attempt it on the sheet.

Up the hill (p15)

Early Learning Goal: Hear and say initial and final sounds in words, and short vowel sounds within words.

What to do: Ask the children to think of words that begin with the 'h' sound. Write them in a list. Ask older children to make groups of 'h' words based on the vowels that follow the 'h' (such as hat, hammer and Harry).

Cool as ice (p16)

Early Learning Goal: Use their phonic knowledge to write simple regular words and make phonetically plausible attempts at more complex words.

What to do: Help the children to complete the sheet. Invite them to use their phonic knowledge to attempt to write the flavour of the ice-cream underneath the cone.

Decorate a jacket (p17)

Early Learning Goal: Use a pencil and hold it effectively to form recognisable letters, most of which are correctly formed.

What to do: Encourage the children to use lots of pre-writing skills to decorate the jacket, such as drawing circles, lines, curves and patterns. Practise air-writing the 'j' before they attempt it on the sheet.

Fly a kite (p18)

Early Learning Goal: Link sounds to letters, naming and sounding the letters of the alphabet.

What to do: Place Post-it notes over the words next to the kites on the sheet. Ask the children to look at the pictures and write the first letter of each word on the corresponding Post-it. How many begin with a 'k'?

Up the ladder (p19)

Early Learning Goal: Use a pencil and hold it effectively to form recognisable letters, most of which are correctly formed.

What to do: Encourage the children to follow the lines carefully, in one colour, without going over the edges. Can they trace another line, in a different colour, along the same route?

Monkey business (p20)

Early Learning Goal: Hear and say initial and final sounds in words, and short vowel sounds within words.

What to do: Set up a sound table in your setting, featuring a different letter sound each week. For the 'm' sound table include objects such as a magazine, a toy mouse, a mango, a muffin and the completed sheets.

Number in the sand (p21)

Early Learning Goal: Use a pencil and hold it effectively to form recognisable letters, most of which are correctly formed.

What to do: Fill some seed trays with damp sand and let the children draw patterns with their fingers. Progress to drawing letters and names in the same way before joining the dots on the sheet.

Playing by the ocean (p22)

Early Learning Goal: Link sounds to letters, naming and sounding the letters of the alphabet.

What to do: Talk to older children about the two phonemes that are represented by the letter 'o' on this page: /o/ (octopus, orange) and /oe/ (ocean). Discuss also how the letters 'oa' (oar) represent the phoneme /or/. Make two lists together of several words that use the /o/ or the /oe/ phonemes.

Puzzle pieces (p23)

Early Learning Goal: Extend their vocabulary, exploring the meanings and sounds of new words.

What to do: Photocopy the children's completed sheets and ask them to cut out the pictures. Make a simple book for each child and suggest that they stick the pictures into them, adding any other 'p' words that they can.

Quiet please! (p24)

Early Learning Goal: Link sounds to letters, naming and sounding the letters of the alphabet.

What to do: Look through a picture dictionary with the children. Notice how the letter 'q' always has a 'u' beside it. Make up some funny sentences together, such as 'Ask quick, quiet questions under the quilt!'. Show the children the lower case and capital 'q's on the sheet.

The end of the rainbow (p25)

Early Learning Goal: Enjoy listening to and using spoken and written language, and readily turn to it in their play and learning.

What to do: What treasure is at the end of Clifford's rainbow? Ask each child to draw their own rainbow. Can they think of some other treasures beginning with 'r' to go at the end of their rainbows?

Sorting socks (p26)

Early Learning Goal: Attempt writing for different purposes, using features of different forms such as lists, stories and instructions.

What to do: Cut out the children's sun pictures from the sheet and stick them onto the centre of an A4 piece of paper. Ask the children to draw sunny day pictures beginning with 's' around the outside (sunglasses, sea, sandcastles etc). Ask older children to attempt to label each 's' picture.

Tidy the toys (p27)

Early Learning Goal: Hear and say initial and final sounds in words, and short vowel sounds within words.

What to do: Make sound posting boxes by cutting slits into four boxes and fixing a letter onto each one. Cut out pictures from magazines that correspond to the four letters. Ask the children to post them in the correct boxes. Now complete the sheet.

It's raining! (p28)

Early Learning Goal: Use a pencil and hold it effectively to form recognisable letters, most of which are correctly formed.

What to do: Practise drawing curvy lines in the air and on paper before completing this sheet. Encourage the children to use curves rather than straight lines to join the dots.

Be my valentine (p29)

Early Learning Goal: Use a pencil and hold it effectively to form recognisable letters, most of which are correctly formed.

What to do: Encourage the children to keep their line within the vine as they draw. Use coloured pencils to make wiggly patterns to decorate the heart.

Watching the waves (p30)

Early Learning Goal: Use their phonic knowledge to write simple regular words and make phonetically plausible attempts at more complex words.

What to do: Complete the sheet with the children, reading all the words and exaggerating the initial sounds. Help older children to think of other 'w' words to add to the sheet, attempting the spellings themselves.

X-ray pictures (p31)

Early Learning Goal: Write their own names and other things such as labels and captions and begin to form simple sentences, sometimes using punctuation.

What to do: Complete the sheet together and extend the activity by helping each child to make an 'X marks the spot' treasure map picture. Encourage older children to add labels to their maps.

Yellow yo-yos (p32)

Early Learning Goal: Hear and say initial and final sounds in words, and short vowel sounds within words.

What to do: What other 'y' words can the children think of? Look through a picture dictionary together to find some more. Ask the children to hide another 'y' thing in Clifford's garden. Do some more work on alliteration using colour names such as 'red raspberries', 'blue balloons' and so on.

Peepo! (p33)

Early Learning Goal: Use a pencil and hold it effectively to form recognisable letters, most of which are correctly formed.

What to do: Make sure that you work with the children as they complete this sheet. Say the name and letter sound of each letter as you join the dots slowly! Practise writing the letter 'z' in the air before attempting it on paper.

Clifford's new toy (p34)

Early Learning Goal: Attempt writing for different purposes, using features of different forms such as lists, stories and instructions.

What to do: Complete the sheet together, then cut out the picture of the ball. Ask the children to draw some other toys for Clifford to choose from. Cut them out and stick them in home-made books entitled, 'A catalogue for Clifford'.

What's your letter? (p35)

Early Learning Goal: Write their own names and other things such as labels and captions and begin to form simple sentences, sometimes using punctuation.

What to do: Talk to the children about the names on the sheet. Discuss how each name starts with a capital letter. Support each child as they write their name, showing them how to use the writing guide lines.

Trace the letters (p36)

Early Learning Goal: Use a pencil and hold it effectively to form recognisable letters, most of which are correctly formed.

What to do: Talk about the pictures together, exaggerating the initial sounds. After each sound is made, help the children to trace the letters in the air, before tracing over them on the sheet.

One present (p37)

Early Learning Goal: Recognise numerals 1 to 9.

What to do: Work with individual children and use the sheet to find out which numbers they recognise. As each child joins the dots from 1 to 10, say the numbers quietly, encouraging them to join in.

T-Bone's toys (p38)

Early Learning Goal: Count reliably up to 10 everyday objects.

What to do: Provide a box of ten toys (pairs of some) and ask the children to sort them into pairs or individual toys. Develop counting skills by saying – hand me two dolls; one ball, and so on. Complete the sheet together.

Finding Clifford (p39)

Early Learning Goal: Use developing mathematical ideas and methods to solve practical problems.

What to do: Talk about the sheet together. For example, ask the children to point to the road with only one log, or to point to the road with the most logs. Do all the roads lead to Clifford? Which is the quickest route? Why?

Clifford's kite (p40)

Early Learning Goal: Recognise numerals 1 to 9.

What to do: Look at the sheet together. Ask the children to read all the numbers on the kite. Which number appears more than once? Can the children find the same number in the numberline at the bottom of the page?

Five buttons (p41)

Early Learning Goal: Count reliably up to 10 everyday objects.

What to do: Provide a box of buttons for the children to sort. Then ask each child to place five buttons on Clifford's coat before drawing them on. Suggest that they place a button over each one they have drawn, counting again to check the number.

Review 1–5 (p42)

Early Learning Goal: Find one more or one less than a number from one to 10.

What to do: Work with five children at a time. Provide some shells and ask each child to count out five. Ask the first child to place one shell in the middle, the next to place one more, and so on until five. Vary the game by starting with different numbers. Complete the sheet together.

Counting fish (p43)

Early Learning Goal: In practical activities and discussion begin to use the vocabulary involved in adding and subtracting.

What to do: Work with the children as they complete the sheet. Use the opportunity to develop mathematical vocabulary, such as 'how many?'; 'altogether'; 'more'; 'same as' and so on.

Finding bones (p44)

Early Learning Goal: Count reliably up to 10 everyday objects.

What to do: Work closely with younger children and encourage them to start by drawing a circle around each bone. Next, help them to point carefully to each bone as they count.

How many letters? (p45)

Early Learning Goal: Use language such as 'more' or 'less' to compare two numbers.

What to do: Ask the children to write their own name on the sheet before they begin. Encourage them to point to each letter as they count the names. Then compare the two numbers. Which name has more letters? Now ask the children to compare names with a friend.

Count the shapes (p46)

Early Learning Goal: In practical activities and discussion begin to use the vocabulary involved in adding and subtracting.

What to do: Before attempting this sheet practise working practically with objects, adding two and then three groups of objects together. As you add groups together use vocabulary such as, 'how many', 'altogether' and 'add'.

Chasing butterflies (p47)

Early Learning Goal: Begin to relate addition to combining two groups of objects and subtraction to 'taking away'.

What to do: Demonstrate how to add two groups together by counting on from starting points other than one. For example, make two groups of objects – count one group (say 5) and then add the groups together by counting on from '5 – 6,7,8 altogether'. Encourage the children to use this technique to complete the sheet.

Review 6–10 (p48)

Early Learning Goal: Find one more or one less than a number from one to 10.

What to do: Work with the children as they complete this sheet. Suggest that they make a mark inside each balloon as they count, so as not to count the same balloon twice. Compare the groups of balloons, asking questions such as: 'Which group of balloons has one more/less than this group?'. And so on.

Follow the path (p49)

Early Learning Goal: Use developing mathematical ideas and methods to solve practical problems.

What to do: Encourage the children's problem-solving skills with this sheet. Can they identify the longest path without counting? Can they guess how many stones are in the other paths once they know the number of stones in the longest path?

Super scoops! (p50)

Early Learning Goal: In practical activities and discussion begin to use the vocabulary involved in adding and subtracting.

What to do: As the children count the number of scoops on each cone encourage them to say the number sentences: 'three scoops and one more makes four scoops altogether'. Help them to count the number of scoops in all three ice-creams by using counting on skills (see, 'Chasing butterflies' above).

■ SCHOLASTIC
PHOTOCOPIABLE

Presents galore! (p51)

Early Learning Goal: Say and use number names in order in familiar contexts.

What to do: Use a numberline for some counting practice before helping the children to complete the sheet. Play different games using the numberline – for example, count on from different starting points; cover up some numbers and ask the children to work out which numbers are missing.

Juggling fun (p52)

Early Learning Goal: In practical activities and discussion begin to use the vocabulary involved in adding and subtracting.

What to do: Use this sheet with older children, making sure that they are familiar with the addition and equals symbols before they complete it. Practise using the notation with small groups as they count out objects such as counters or cubes. Show them how to record what they are doing, saying the number sentences as you write down the symbols.

Match the puzzles (p53)

Early Learning Goal: Use language such as 'circle' or 'bigger' to describe the shape and size of solids and flat shapes.

What to do: Look at the different shapes in the puzzles on the sheet. Ask the children to describe the shapes using words such as 'curvy', 'sides', 'straight' and so on. Do they know the names of any of the shapes? Challenge them to make a shape puzzle of their own by drawing round some plastic shapes.

Review 11–15 (p54)

Early Learning Goal: Use language such as 'more' or 'less' to compare two numbers.

What to do: Let the children complete the sheet and then cut out each row of stars and each number. Mix the rows and numbers up and play games with them. For example, ask the children to: match the rows of stars with the numbers; choose a row and then find a row that has one more or one less star; add an extra star to a row and then find two rows the same.

Up in the clouds (p55)

Early Learning Goal: Say and use number names in order in familiar contexts.

What to do: Work with the children as they complete this sheet to ensure that they are following and counting the numbers in order. Challenge them to make another weather dot-to-dot picture on the reverse of the sheet for a friend to complete.

Missing numbers (p56)

Early Learning Goal: Use developing mathematical ideas and methods to solve practical problems.

What to do: Before attempting this sheet, the children will need to recognise the numbers from 10 to 20. Practise counting from points other than one, challenging them to say the next number in a random sequence.

Match the shells (p57)

Early Learning Goal: Count reliably up to 10 everyday objects.

What to do: Let the children complete the sheet with minimum guidance if possible. Encourage them to carefully point to each shell as they say the numbers. Continue the fun by sorting shells in the sand tray, counting the number of curvy, spiral, pointy shells and so on.

Starry sky (p58)

Early Learning Goal: In practical activities and discussion begin to use the vocabulary involved in adding and subtracting.

What to do: Complete the sheet with the children. Encourage them to use counting on skills to count all of the stars in the sky. Explain that they know that they have 10 stars, and ask them what number comes next. Demonstrate how to start counting on from 10 saying, 'eleven, twelve …'.

Falling leaves (p59)

Early Learning Goal: Begin to relate addition to combining two groups of objects and subtraction to 'taking away'.

What to do: Complete the sheet with the children. Ask them: 'How can you find out how many leaves there are altogether?' Use vocabulary such as 'add together' and suggest that they count on from the first number, rather than counting all the leaves again. Encourage them to talk in number sentences such as 'Ten and ten more makes 20'.

Complete the puzzle (p60)

Early Learning Goal: Use developing mathematical ideas and methods to solve practical problems.

What to do: Talk about the puzzle pieces with the children. Which is the biggest number? Which is the smallest number? Where do they think the bigger numbers will go (near the beginning or end of the puzzle)? Can the children spot the pattern of the missing puzzle pieces?

Count the bones (p61)

Early Learning Goal: Count reliably up to 10 everyday objects.

What to do: Draw the children's attention to the fact that although each pile of bones looks different there is the same number in each one. Provide them with some sorting objects and challenge them to make four piles of five, arranging each pile differently. Older children should complete the sheet, counting carefully beyond 10 to 20.

Writing numbers (1) and (2) (p62–63)

Early Learning Goal: Say and use number names in order in familiar contexts.

What to do: Help the children to carefully colour in and write the numbers from 1 to 20. Ask them to cut out the numbers in strips and then put them back in the correct order. Let each child fix the numbers (using glue sticks) to a long strip of card to make their own personal numberline.

Drop a penny (p64)

Early Learning Goal: Recognise numerals 1 to 9.

What to do: Enjoy this game together. After you have played a few times make up or play some different number games. Ideas include rolling pennies onto a grid of numbers, hopscotch, and track and board games such as Lotto and Snakes and Ladders.

Circle the pictures

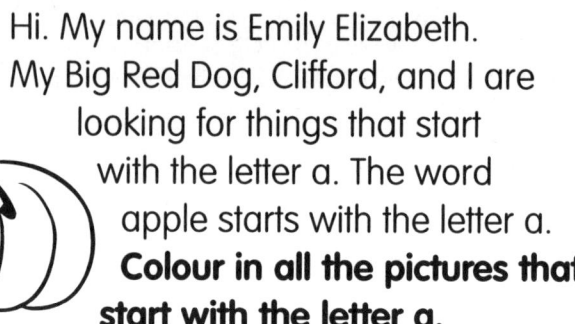

Hi. My name is Emily Elizabeth. My Big Red Dog, Clifford, and I are looking for things that start with the letter a. The word apple starts with the letter a. **Colour in all the pictures that start with the letter a.**

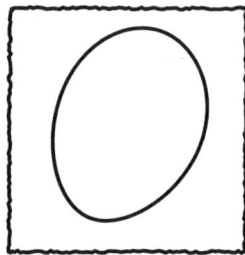

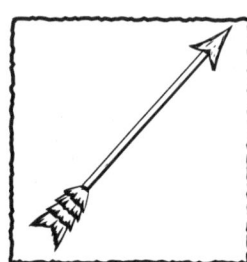

egg   arrow   apple

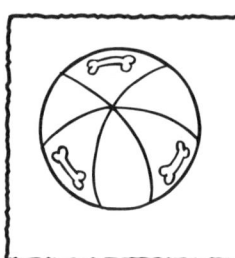

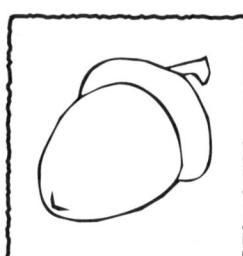

toy   ant   ball   acorn

a b c d e f g h i j k l m n o p q r s t u v w x y z

Bat and ball

T-Bone likes to play with a ball.
The word ball starts with a b.
Everything on this page starts with
the letter b.
Write the first letter of each word.

bat

alloon

all

ee

**Draw something else
that starts with the
letter b here.**

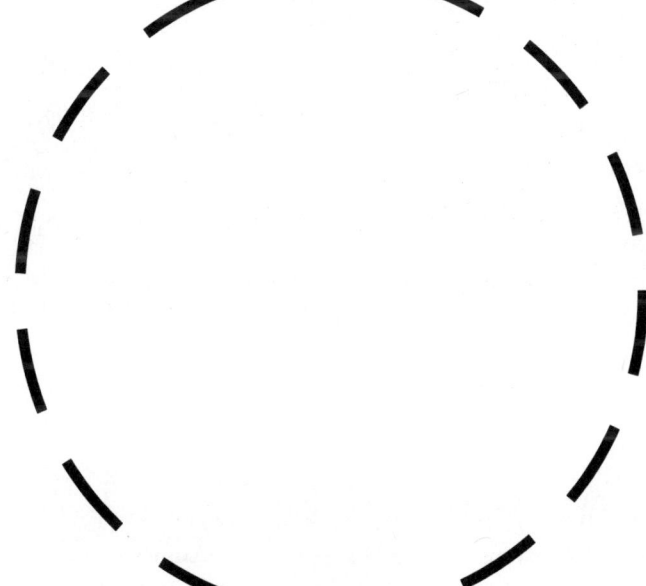

ird

a **b** c d e f g h i j k l m n o p q r s t u v w x y z

SCHOLASTIC
PHOTOCOPIABLE

Clifford and Cleo

Clifford's and Cleo's names both start with the letter C.
Find the thing in each row that does not start with the letter c. Mark an X on it.

cat cow dog

hat carrot cone

cup car pig

Draw something else that starts with the letter c here.

a b **c** d e f g h i j k l m n o p q r s t u v w x y z

■ SCHOLASTIC
PHOTOCOPIABLE

Join the dots

Clifford is a dog. The word dog starts with the letter d.
Join the dots to find something else that starts with the letter d.

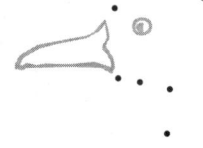

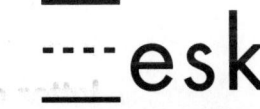

drum   esk   og

These things also start with the letter d. Colour them in.

a b c **d** e f g h i j k l m n o p q r s t u v w x y z

SCHOLASTIC
PHOTOCOPIABLE

The eagle's eggs

My name is Emily Elizabeth. Both of my names start with the letter E.
The words eagle, eggs, and eight also start with the letter e.
Help me follow the road that leads to the eagle's eggs.

eight eagle eggs

**Write how
many
eggs are in
the nest:**  _e_ggs

a b c d **e** f g h i j k l m n o p q r s t u v w x y z

SCHOLASTIC
PHOTOCOPIABLE

Fishing lines

T-Bone is looking at a fish. The word fish starts with the letter f.
Draw lines between the fish that are the same.

frog   ox

These things also start with the letter f. Colour them in.

a b c d e **f** g h i j k l m n o p q r s t u v w x y z

SCHOLASTIC
PHOTOCOPIABLE

Behind the gate

Hi. I'm a girl. The word girl starts with the letter g. The words gate and grass start with the letter g, too. Clifford is behind one of the gates. **Which road leads to Clifford?**

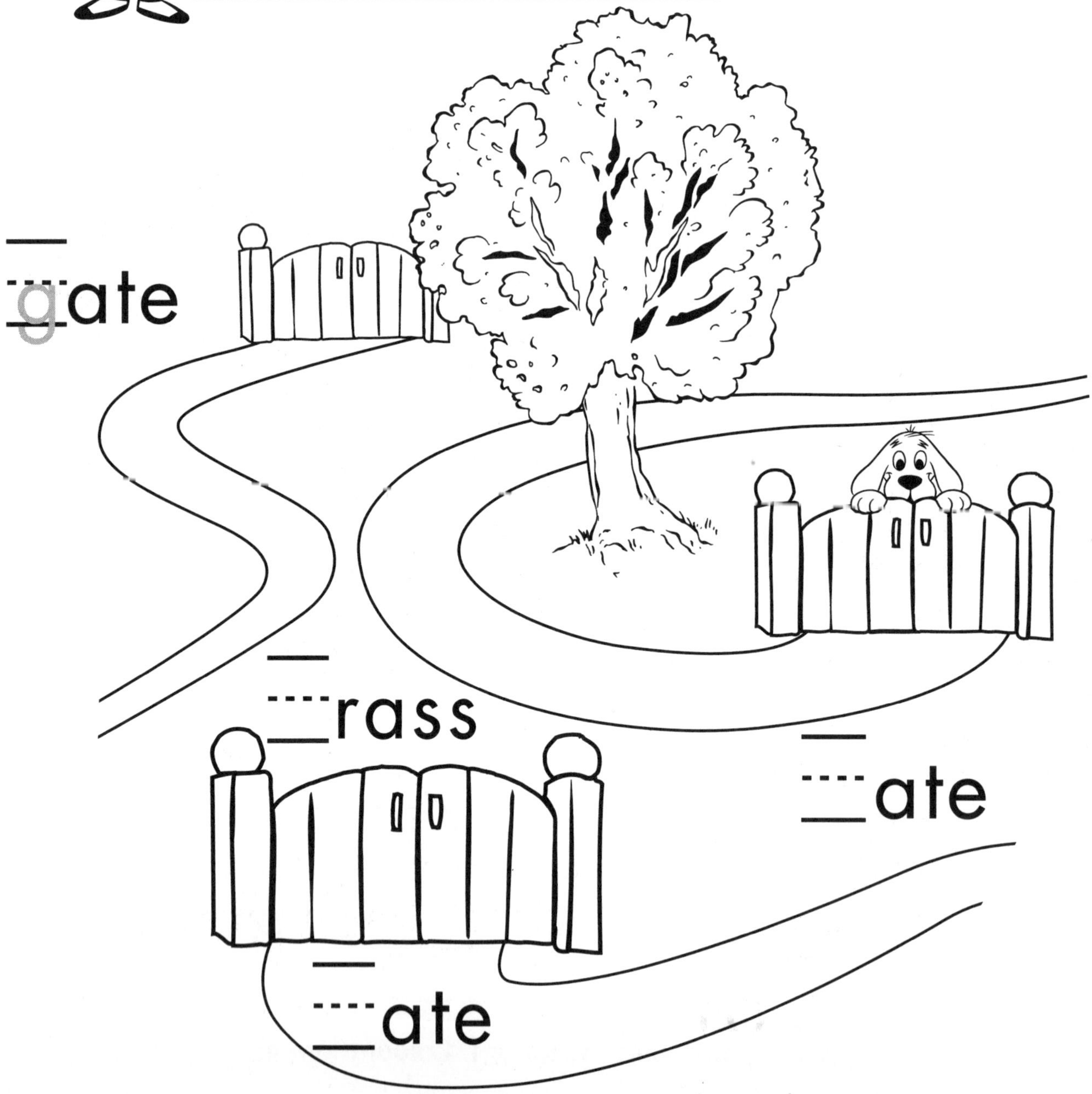

gate

__rass

__ate

__ate

a b c d e f **g** h i j k l m n o p q r s t u v w x y z

■SCHOLASTIC
PHOTOCOPIABLE

Up the hill

Clifford is up on a hill. The word hill starts with the letter h. Some of the things at the bottom of the hill start with the letter h too.

Follow all the roads that will take Clifford down to the h words.
Write the first letter of each word.

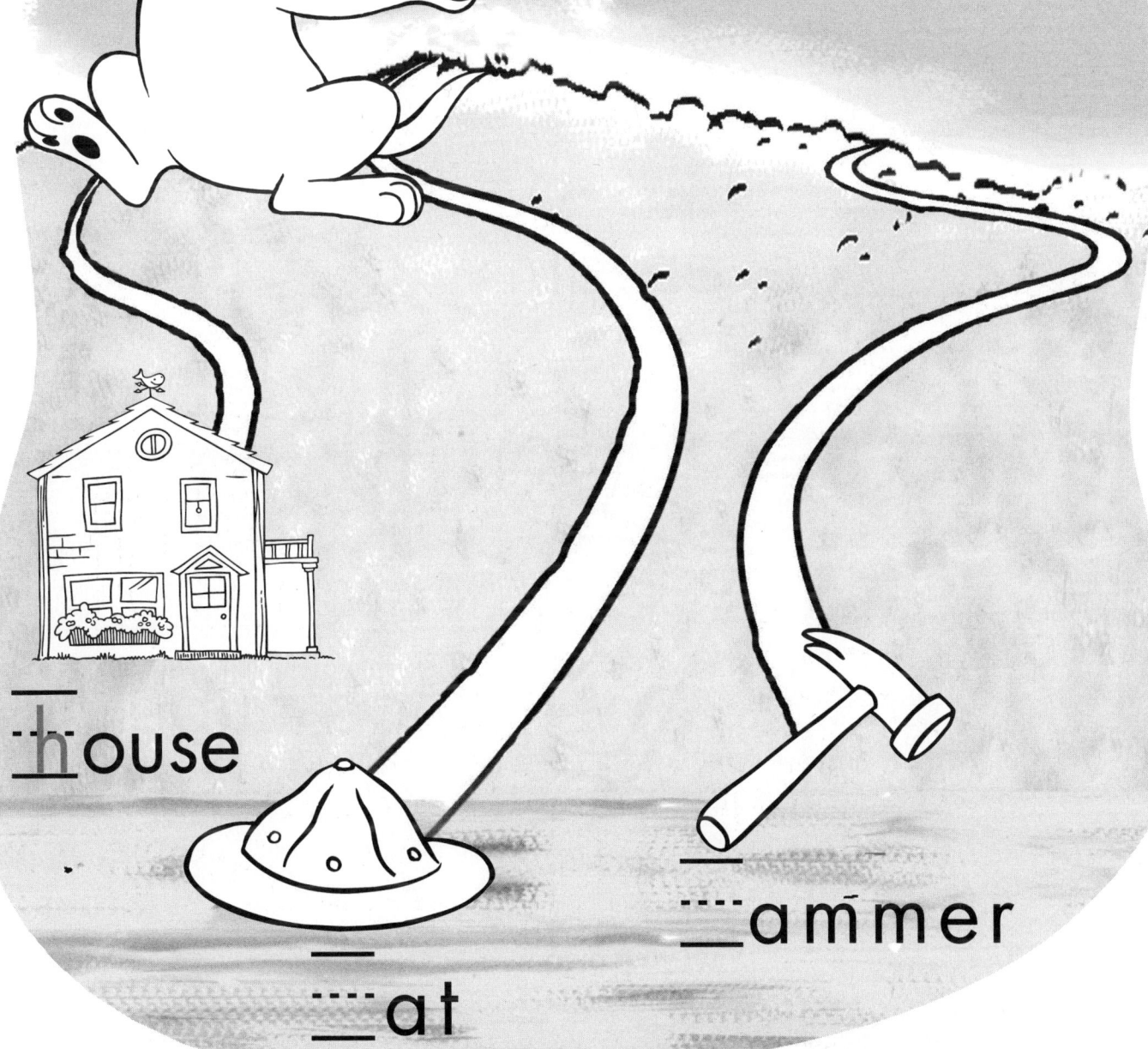

<u>h</u>ouse

<u>h</u>at

<u>h</u>ammer

a b c d e f g **h** i j k l m n o p q r s t u v w x y z

Cool as ice

T-Bone and Cleo really like ice-cream! They also like lemonade with ice-cubes. The word ice starts with the letter i.

Colour the ice-cream cone.

Fill in the missing letters.

ice cream

ice cube

Draw your own ice-cream cone here.

a b c d e f g h **i** j k l m n o p q r s t u v w x y z

SCHOLASTIC
PHOTOCOPIABLE

Decorate a jacket

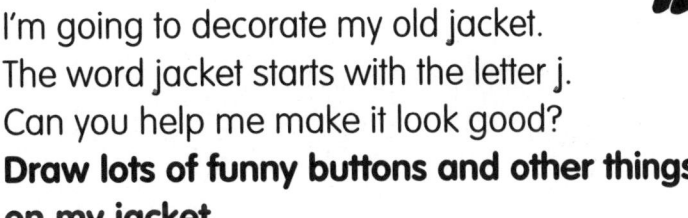

I'm going to decorate my old jacket.
The word jacket starts with the letter j.
Can you help me make it look good?
Draw lots of funny buttons and other things on my jacket.

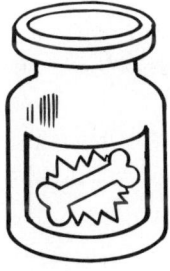 __jar    __umbo __et    __am

These things also start with the letter j.

a b c d e f g h i **j** k l m n o p q r s t u v w x y z

SCHOLASTIC
PHOTOCOPIABLE

Foundation Stage

Fly a kite

T-Bone is dreaming that he can fly like a kite. The word kite starts with the letter k. Let's find some other things that start with k.
Colour the kites with k words.

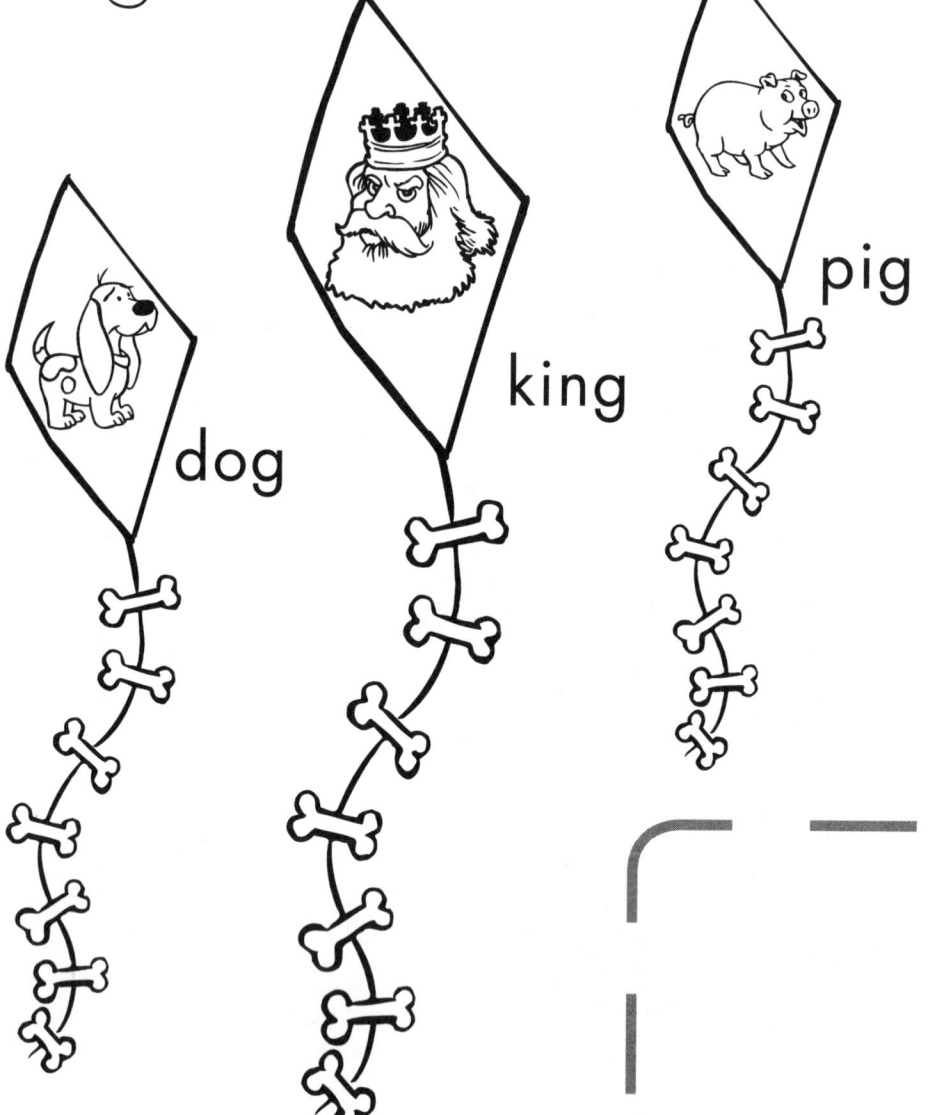

dog

king

pig

key

Draw something else that starts with the letter k here.

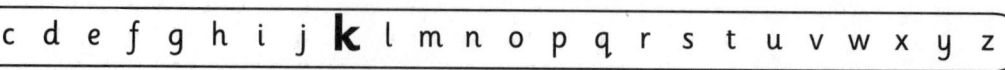

■ SCHOLASTIC
PHOTOCOPIABLE

Up the ladder

Cleo is trying to reach the ladder. The word ladder starts with the letter l.

Follow the road that will take Cleo to the ladder. What is at the top of the ladder?

lunch box

‎‑‑‑‑ladder

‑‑‑‑eaf

‑‑‑‑og

a b c d e f g h i j k **l** m n o p q r s t u v w x y z

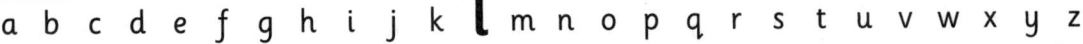

Monkey business

There are 4 monkeys and 4 mice hiding on this page. The words monkey and mice start with the letter m.

Can you help me to find and circle them?

mice ____onkey

Clifford's First Letters and Numbers

SCHOLASTIC
PHOTOCOPIABLE

Number in the sand

Clifford has found a number written in the sand. The word number starts with the letter n.
Can you help Clifford join the dots? What number is in the sand?

___
‾n̲ine

The word nest also starts with the letter n.
Draw a nest here.

___
‾‾‾est

a b c d e f g h i j k l m **n** o p q r s t u v w x y z

SCHOLASTIC
PHOTOCOPIABLE

Foundation Stage

21

Name: Date:

Playing by the ocean

Clifford and I like to play by the ocean. The word ocean starts with the letter o.
Draw circles around the other things on this page that start with the letter o. Write the first letter of each word below.

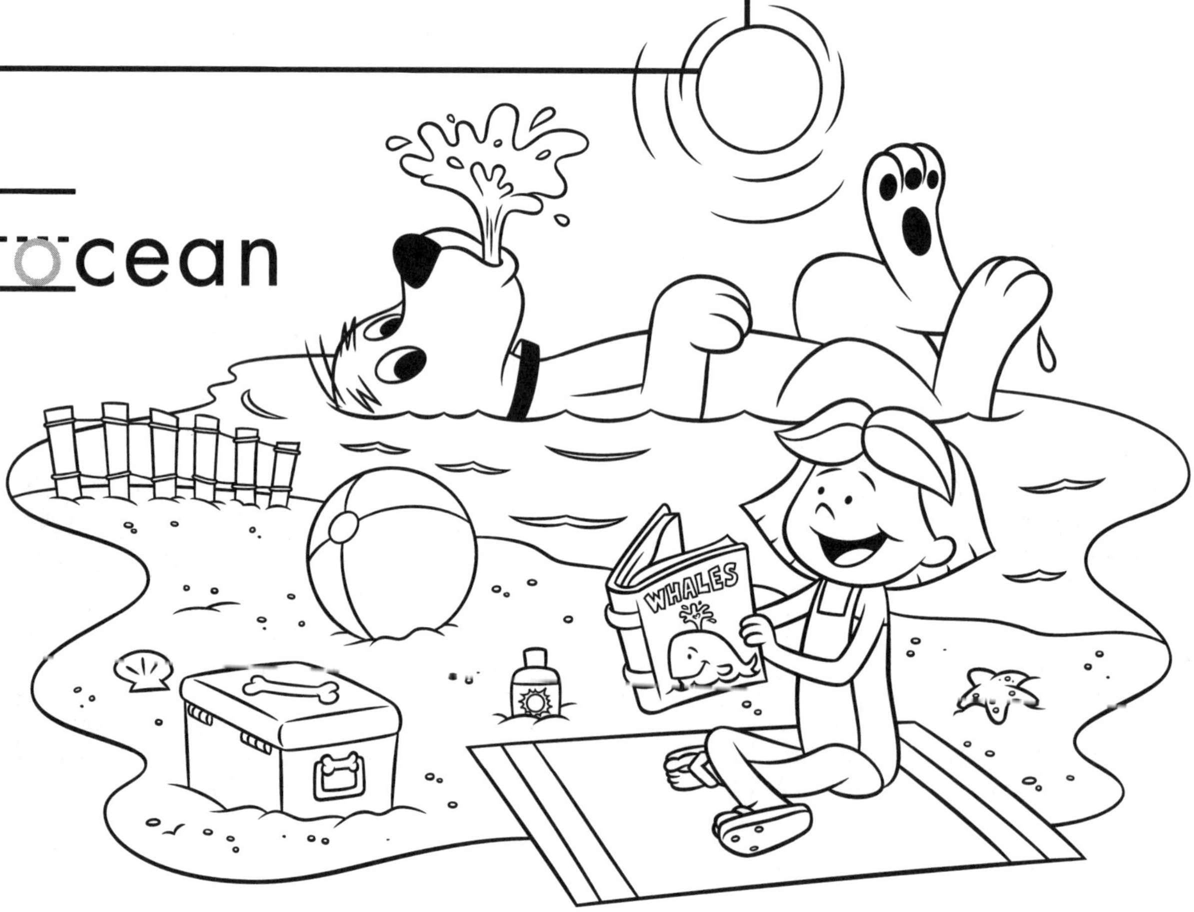

o cean

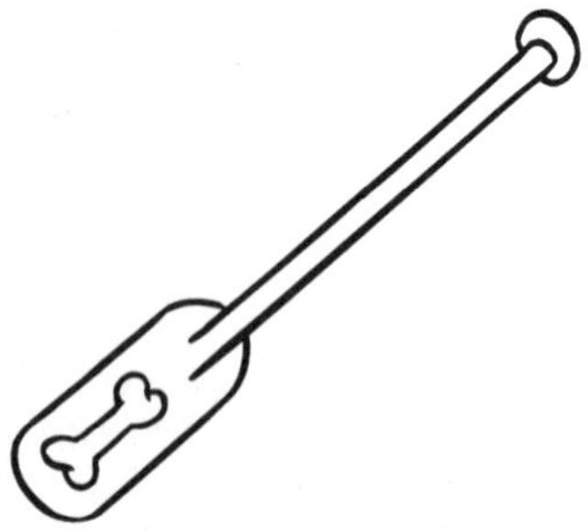

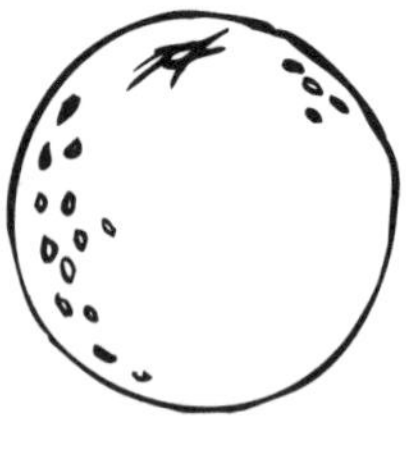

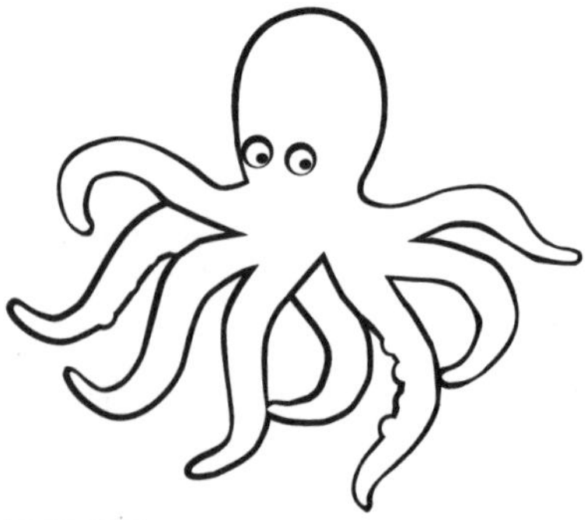

_ar _range _ctopus

a b c d e f g h i j k l m n **o** p q r s t u v w x y z

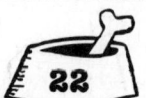

SCHOLASTIC PHOTOCOPIABLE

Puzzle pieces

I'm doing a puzzle. The word puzzle starts with the letter p.
Help me finish my puzzle.
Draw a line from each piece to the place where it belongs.

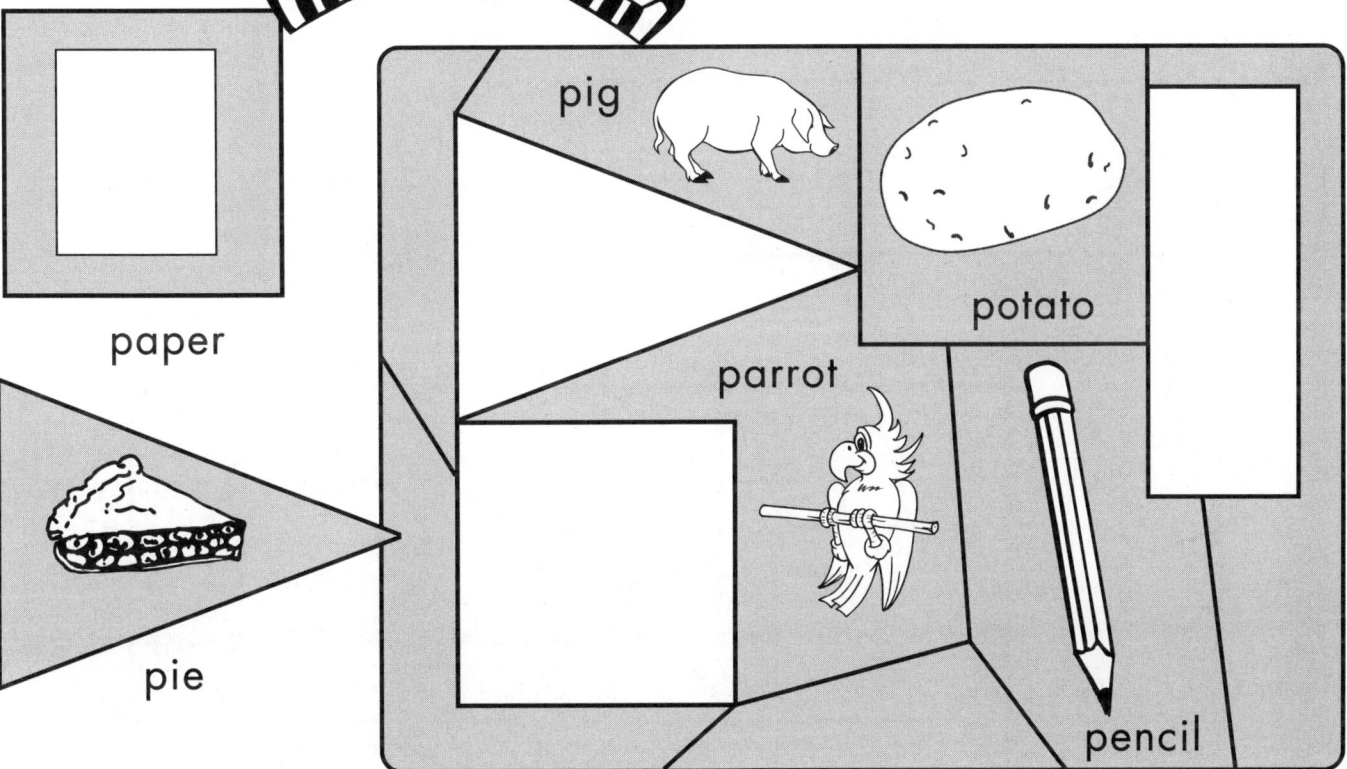

paper

pie

pig

potato

parrot

pencil

pen

Draw something else that starts with the letter p here.

a b c d e f g h i j k l m n o **p** q r s t u v w x y z

Quiet please!

Shhh. BE QUIET! T-Bone
wants to have a sleep on that
quilt! The words quilt and
QUIET start with q.
**How many capital Q's do
you see? How many
lower case q's do you
see?**

QUIET

quilt

a b c d e f g h i j k l m n o p **q** r s t u v w x y z

SCHOLASTIC
PHOTOCOPIABLE

The end of the rainbow

Clifford is red. The word red starts with the letter r. The words rain and rainbow start with an r, too.

Colour the rainbow.
What is at the end of the rainbow?
Say the words out loud.

_ _ rain

_ _ ose

_ _ ug

_ _ ing

a b c d e f g h i j k l m n o p q **r** s t u v w x y z

SCHOLASTIC PHOTOCOPIABLE

Sorting socks

Hey! Let's sort socks! I can see 6 socks on this page. Can you? The words six and socks start with the letter s.

Draw lines between the socks that are the same.

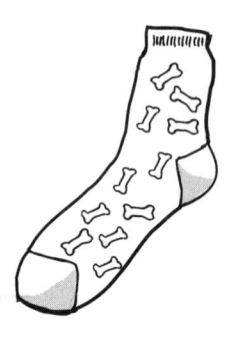
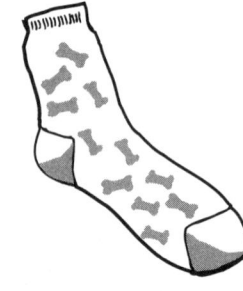
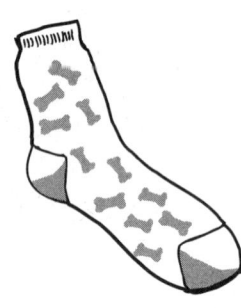
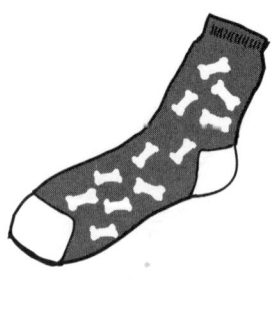

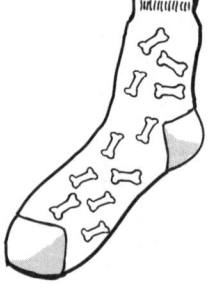

un

The word sun also starts with the letter s. Draw a sun here.

s ocks

a b c d e f g h i j k l m n o p q r **S** t u v w x y z

SCHOLASTIC
PHOTOCOPIABLE

Tidy the toys

T-Bone's name starts with the letter T. So does the word toy! Can you help us put our toys away?
Draw lines from the things that start with t to the toy box.

top

tortoise

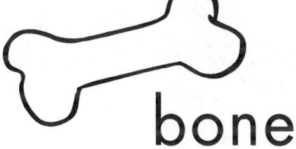

bone

tiger

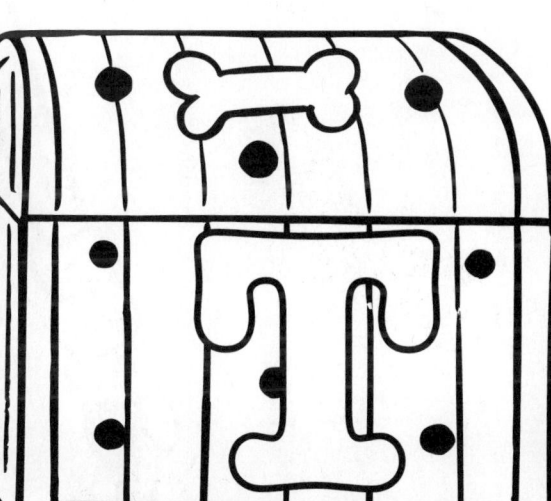

bucket

ball

a b c d e f g h i j k l m n o p q r s **t** u v w x y z

It's raining!

It's raining. I need to put something up to keep me dry. The word up starts with the letter u.
Join the dots to see what I need. It starts with the letter u too.
Colour it in.

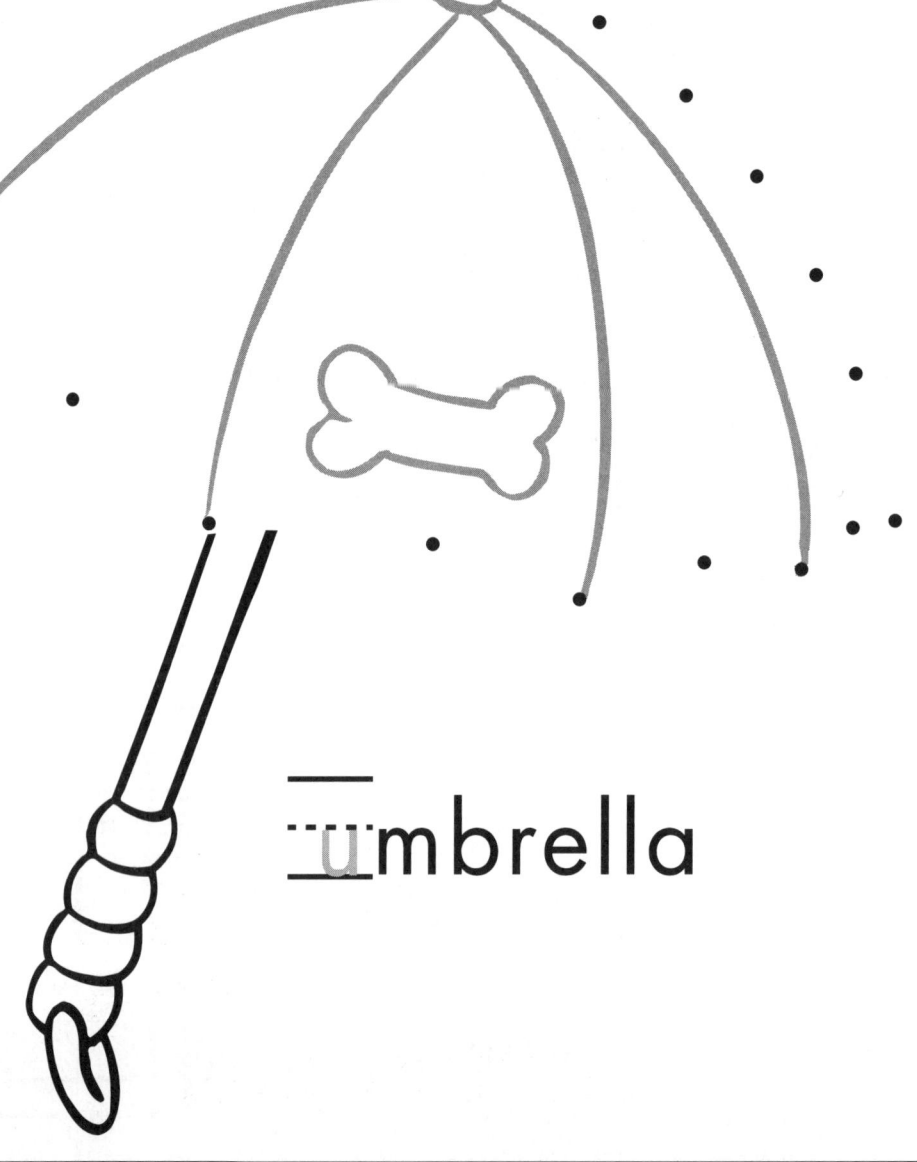

u̲mbrella

Be my valentine

Oh dear! Can you help Cleo get through these vines? The word vine starts with the letter v.
Follow the road that will take Cleo out of the vines. What is at the end of the road? Decorate it.

‗‗‗ine

‗valentine

a b c d e f g h i j k l m n o p q r s t u **v** w x y z

Name: Date:

Watching the waves

Cleo and T-Bone are near the water. They are watching the waves. The word water starts with the letter w. So does the word wave.
Circle all the other things that start with w.
Write the first letter of each word.

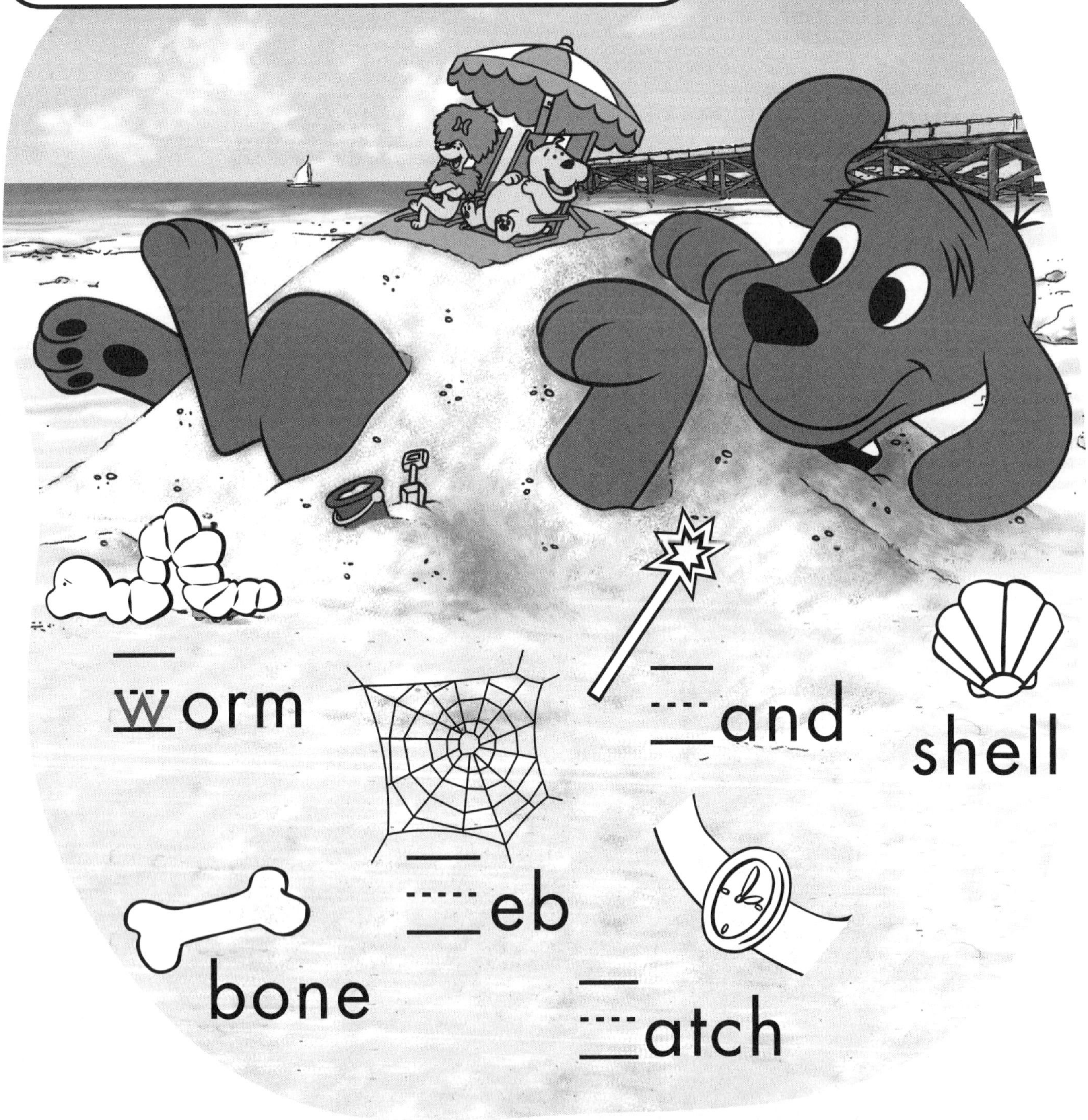

worm
___and
shell
___eb
bone
___atch

a b c d e f g h i j k l m n o p q r s t u v **w** x y z

SCHOLASTIC PHOTOCOPIABLE

X-ray pictures

Look at all these X-rays! X-ray starts with the letter X.

Mark an X on the X-ray in each row that is not like the others.

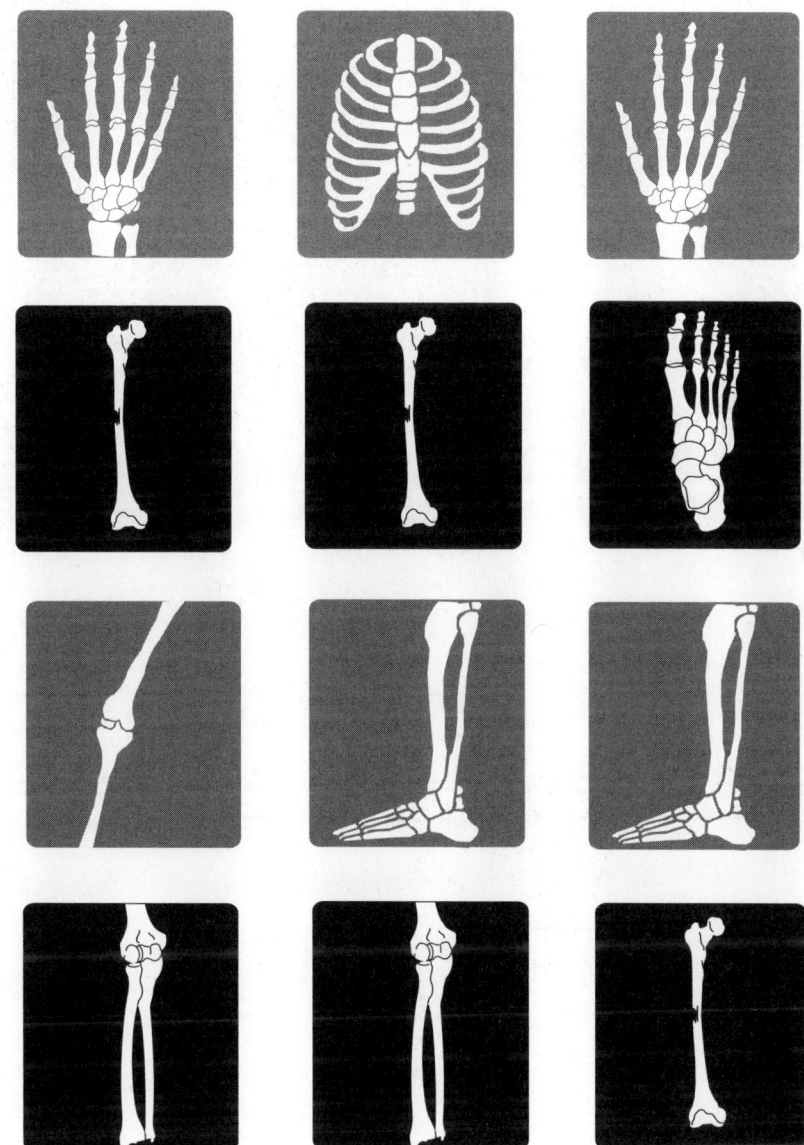

SCHOLASTIC
PHOTOCOPIABLE

Yellow yo-yos

Clifford is looking for yellow yo-yos in his garden. The words yellow and yo-yo start with the letter y.
Colour all the yo-yos yellow.

yellow ─o-─o

a b c d e f g h i j k l m n o p q r s t u v w x **y** z

SCHOLASTIC
PHOTOCOPIABLE

Peepo!

T-Bone is playing peepo with an animal in the zoo. The name of the animal starts with the letter z. So does the word zoo.

Join the dots to see T-Bone's new friend.

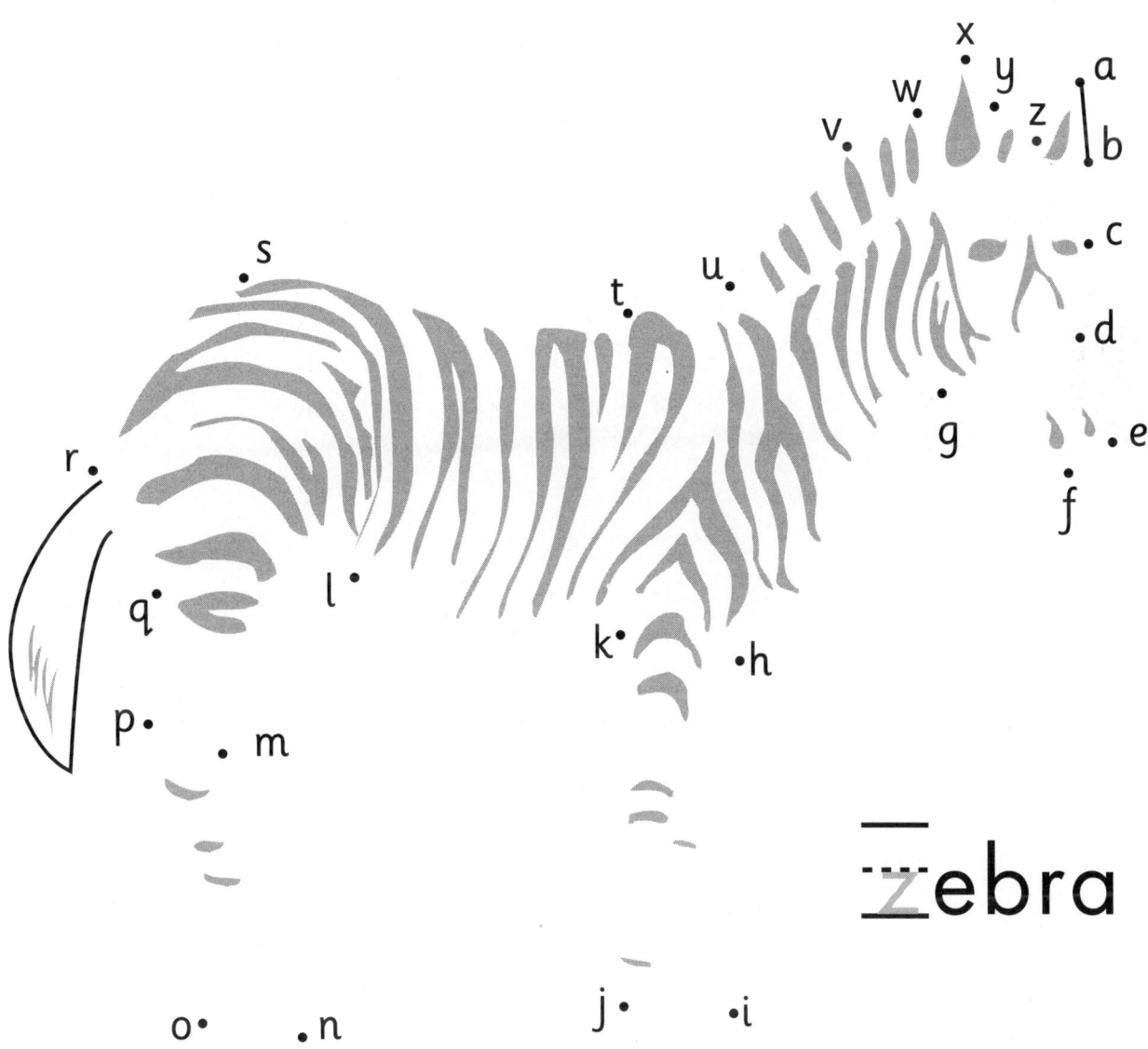

zebra

a b c d e f g h i j k l m n o p q r s t u v w x y **z**

Clifford's new toy

Clifford has a new toy. **Follow the letters from a to z to see his new toy.**

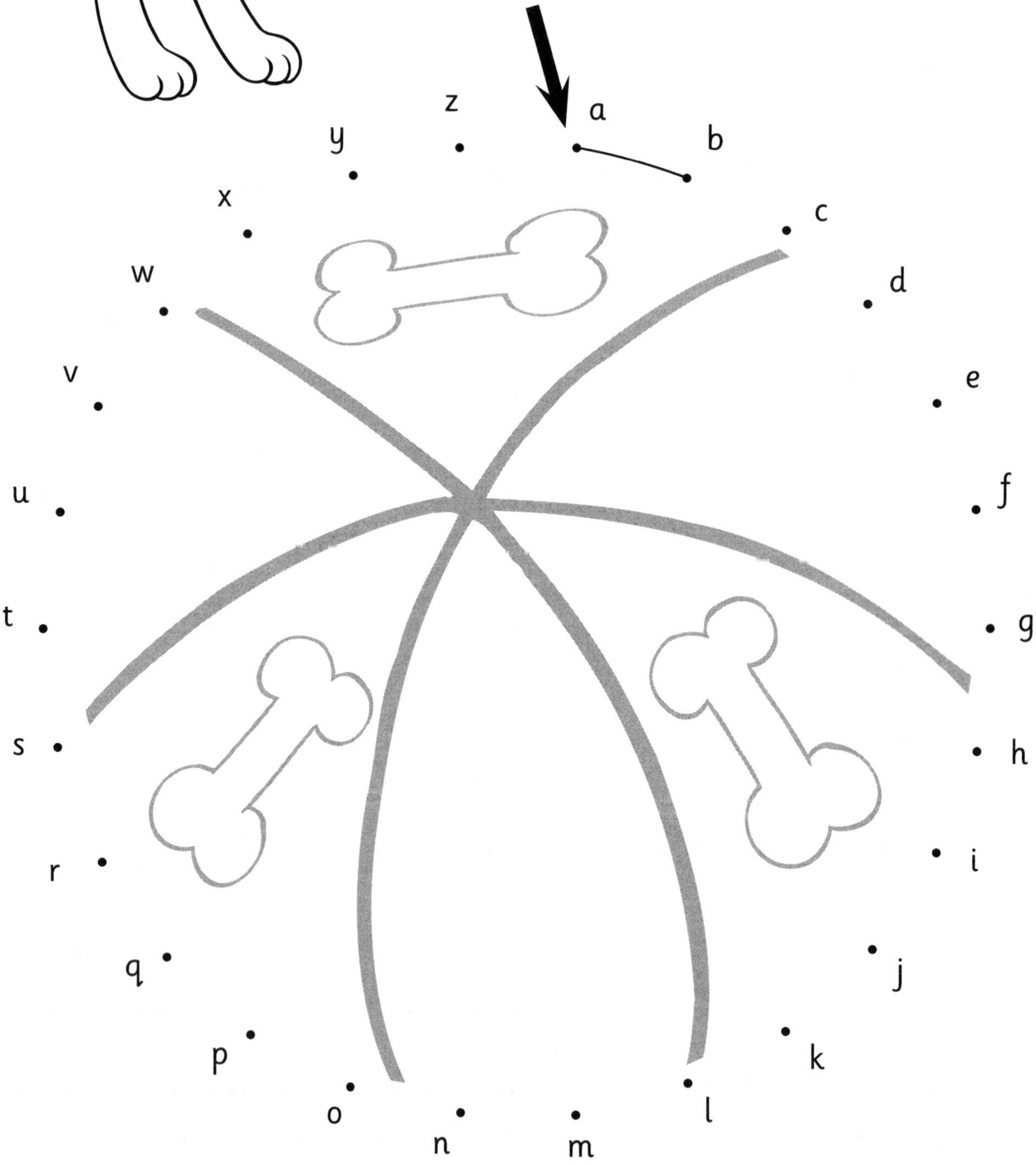

z
y   a   b
x   c
w   d
v   e
u   f
t   g
s   h
r   i
q   j
p   k
o   n   m   l

a b c d e f g h i j k l m n o p q r s t u v w x y z

SCHOLASTIC PHOTOCOPIABLE

What's your letter?

Cleo's and Clifford's names start with the letter C. T-Bone's name starts with the letter T. My name starts with the letter E. What does your name start with?

Write the letter

_____

Draw something that starts with the same letter here.

Write your name here

_____

a b c d e f g h i j k l m n o p q r s t u v w x y z

Trace the letters

Look what Clifford has found.
**Draw lines between the things
that start with the same letter.
Trace the first letter of each
word.**

apple

cap

cup

boot

bone

arrow

a b c d e f g h i j k l m n o p q r s t u v w x y z

SCHOLASTIC
PHOTOCOPIABLE

One present

Hi. My name is Emily Elizabeth. My Big Red Dog, Clifford, is my best friend. Today I gave him 1 present.

Follow the numbers from 1 to 10 to see what I gave him.

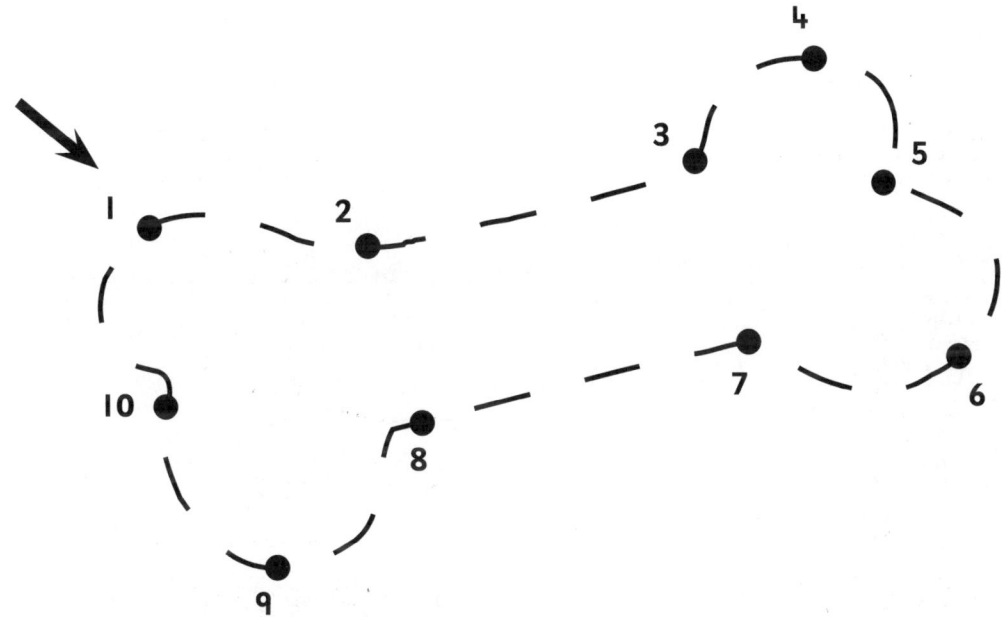

How many bones can you see on this page?

Write the number.   _____

SCHOLASTIC
PHOTOCOPIABLE

T-Bone's toys

Help T-Bone to sort out his toys. He has 2
balls that look just the same.
Draw lines between them.
**Draw lines between the other toys that
look the same.**

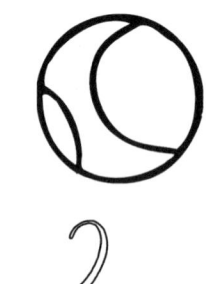

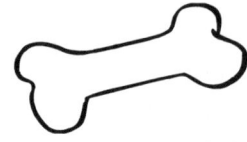

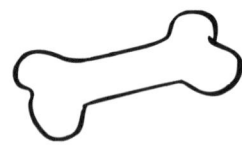

How many balls can you see on this page?

Write the number.   - - - - - - - - - - - - - - -

1 **2** 3 4 5 6 7 8 9 10 11 12 13 14 15 16 17 18 19 20

Finding Clifford

Cleo is looking for Clifford. Can you help her to find him?

Draw a line along the road that has 3 logs on it.

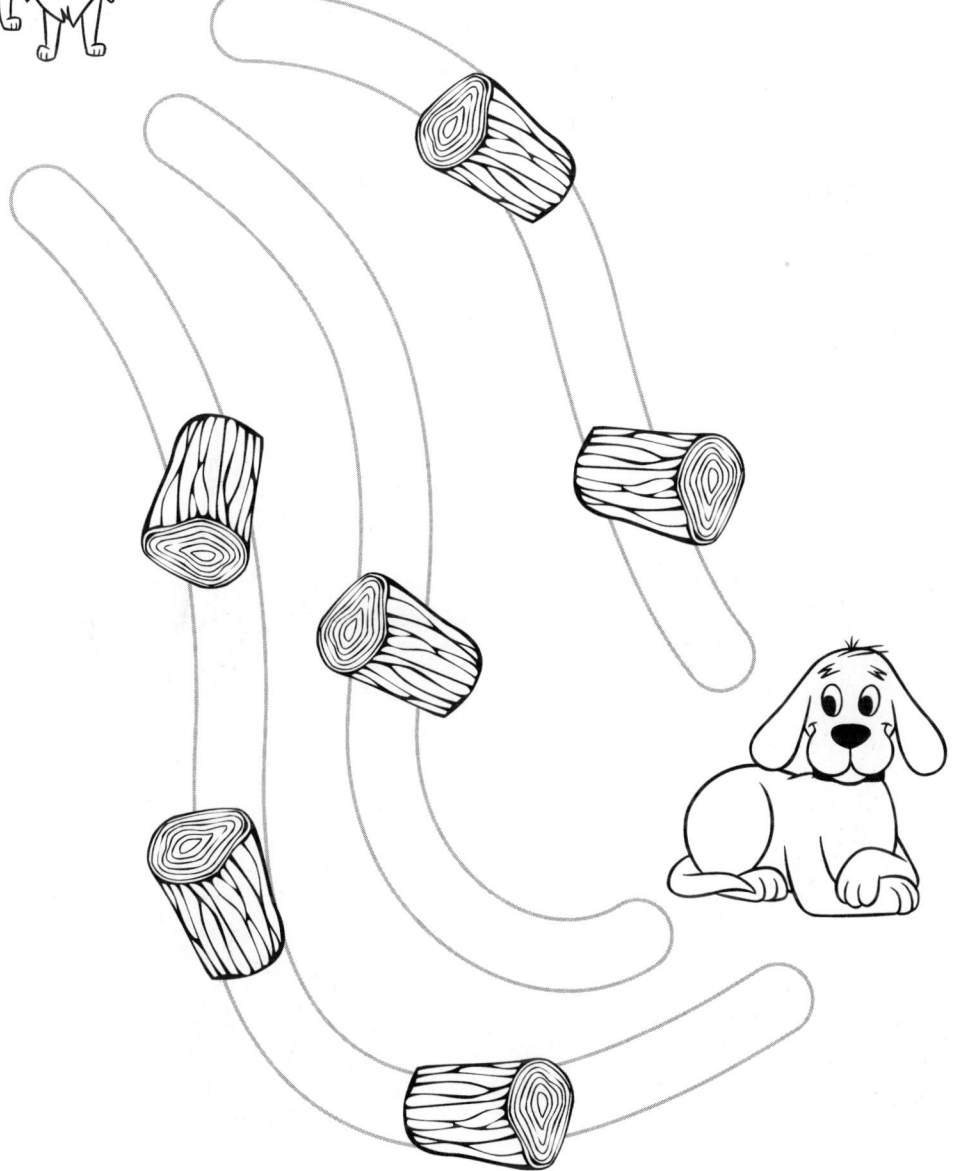

How many different roads can you see?

_____

Write the number.   - - - - - - - - - - - - - -

_____

| 1 | 2 | **3** | 4 | 5 | 6 | 7 | 8 | 9 | 10 | 11 | 12 | 13 | 14 | 15 | 16 | 17 | 18 | 19 | 20 |

Clifford's kite

Clifford and I have a new kite. Help us to decorate it.

Colour all the circles with the number 4 red.
Colour the other circles any colours you like.

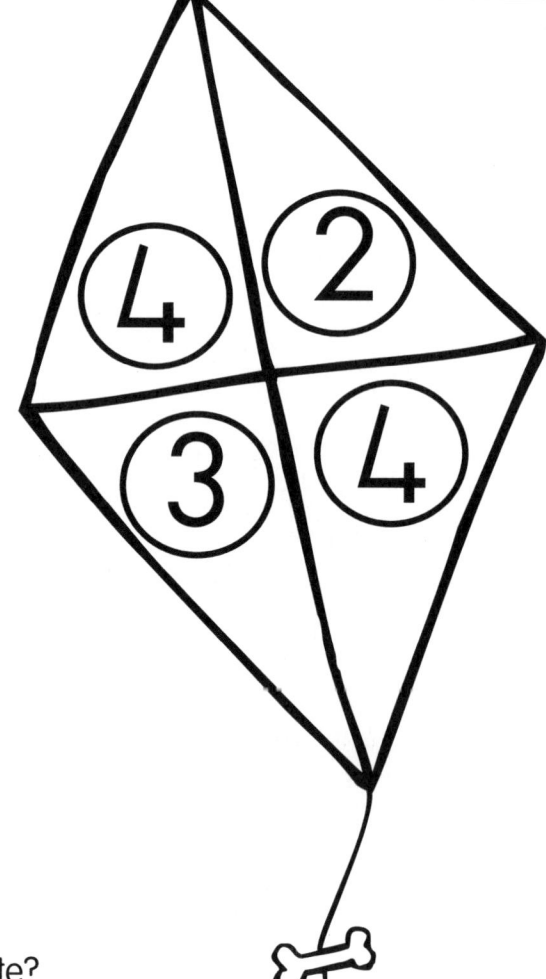

How many circles can you see on the kite?

Write the number. _____

1 2 3 **4** 5 6 7 8 9 10 11 12 13 14 15 16 17 18 19 20

SCHOLASTIC
PHOTOCOPIABLE

Five buttons

Clifford's fur keeps him nice and warm. But I'm making him a special coat for when it gets very, very cold.
Draw 5 buttons on Clifford's new coat.

How many pockets can you see on Clifford's new coat?

_____

Write the number. - - - - - - - - - - - - - - -

_____

| 1 | 2 | 3 | 4 | **5** | 6 | 7 | 8 | 9 | 10 | 11 | 12 | 13 | 14 | 15 | 16 | 17 | 18 | 19 | 20 |

📖 SCHOLASTIC
PHOTOCOPIABLE

Foundation Stage

Review 1–5

Clifford and I are counting. You can help us.
Write one number on each line.

How many cakes?   _____

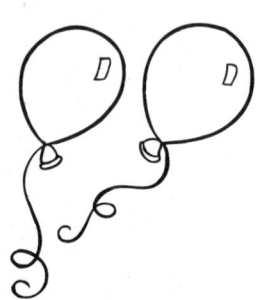

How many balloons?   _____

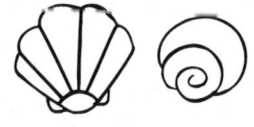

How many shells?   _____

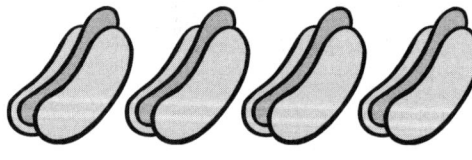

How many hot dogs?   _____

How many candles?   _____

1 2 3 4 5 6 7 8 9 10 11 12 13 14 15 16 17 18 19 20

SCHOLASTIC
PHOTOCOPIABLE

Name: Date:

Counting fish

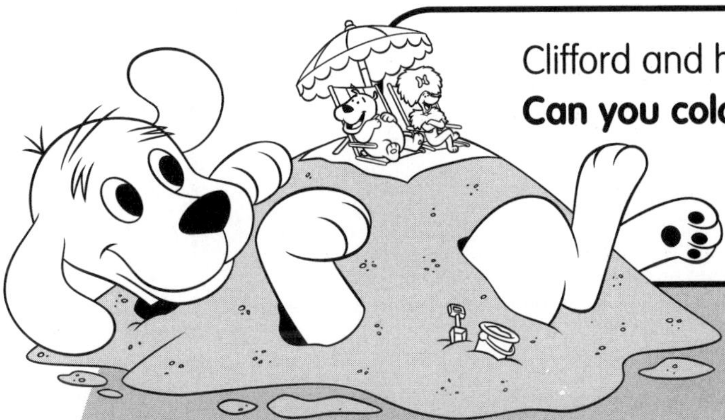

Clifford and his friends are counting fish.

Can you colour 3 fish blue? Circle each of them.

Can you colour 3 fish green?

Mark an X on each green fish.

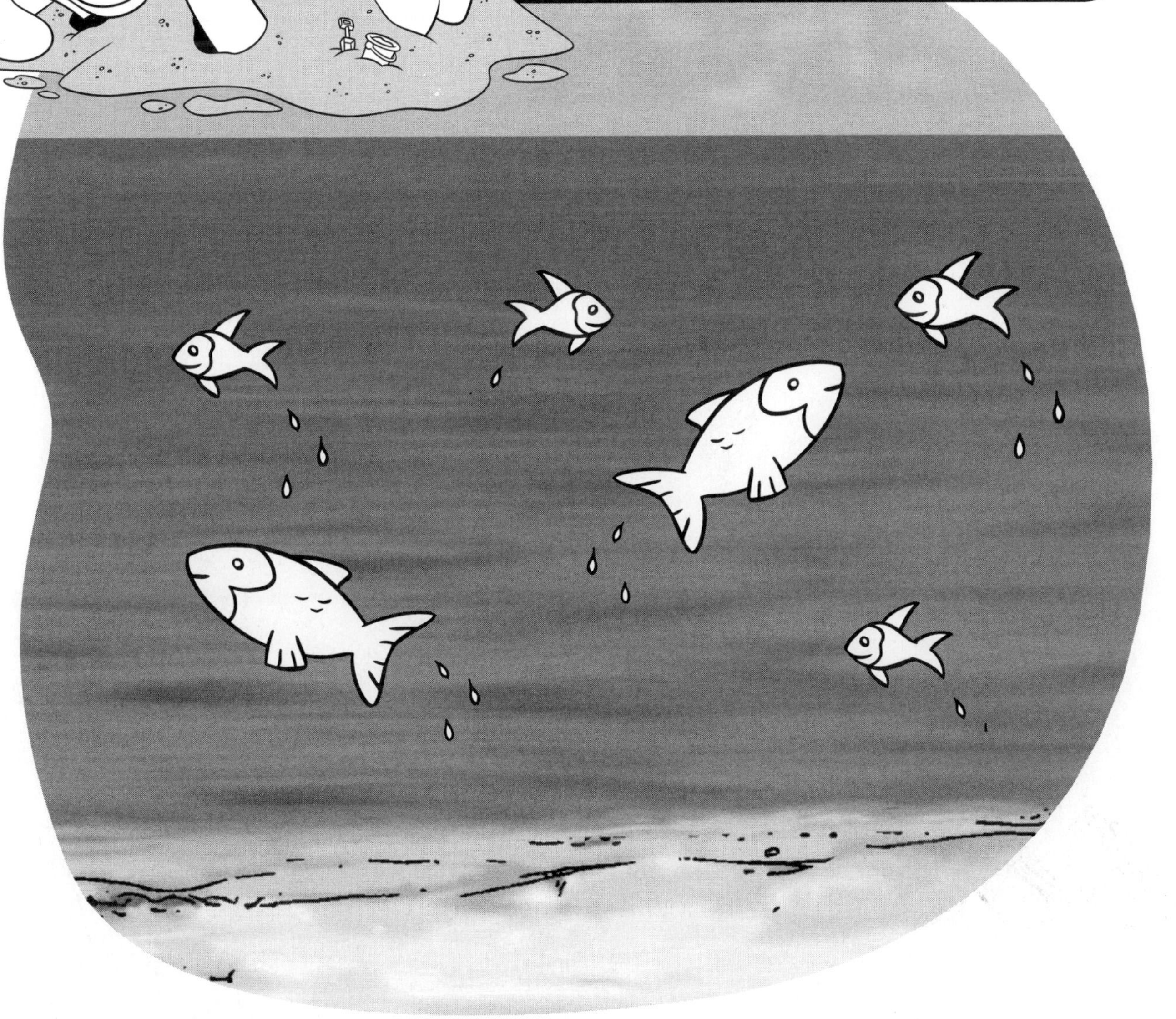

Count all the fish on this page. How many fish can you see?

Write the number.

1 2 3 4 5 **6** 7 8 9 10 11 12 13 14 15 16 17 18 19 20

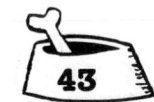

Finding bones

T-Bone is busy hiding bones.
How many bones can you find?
Circle the bones.

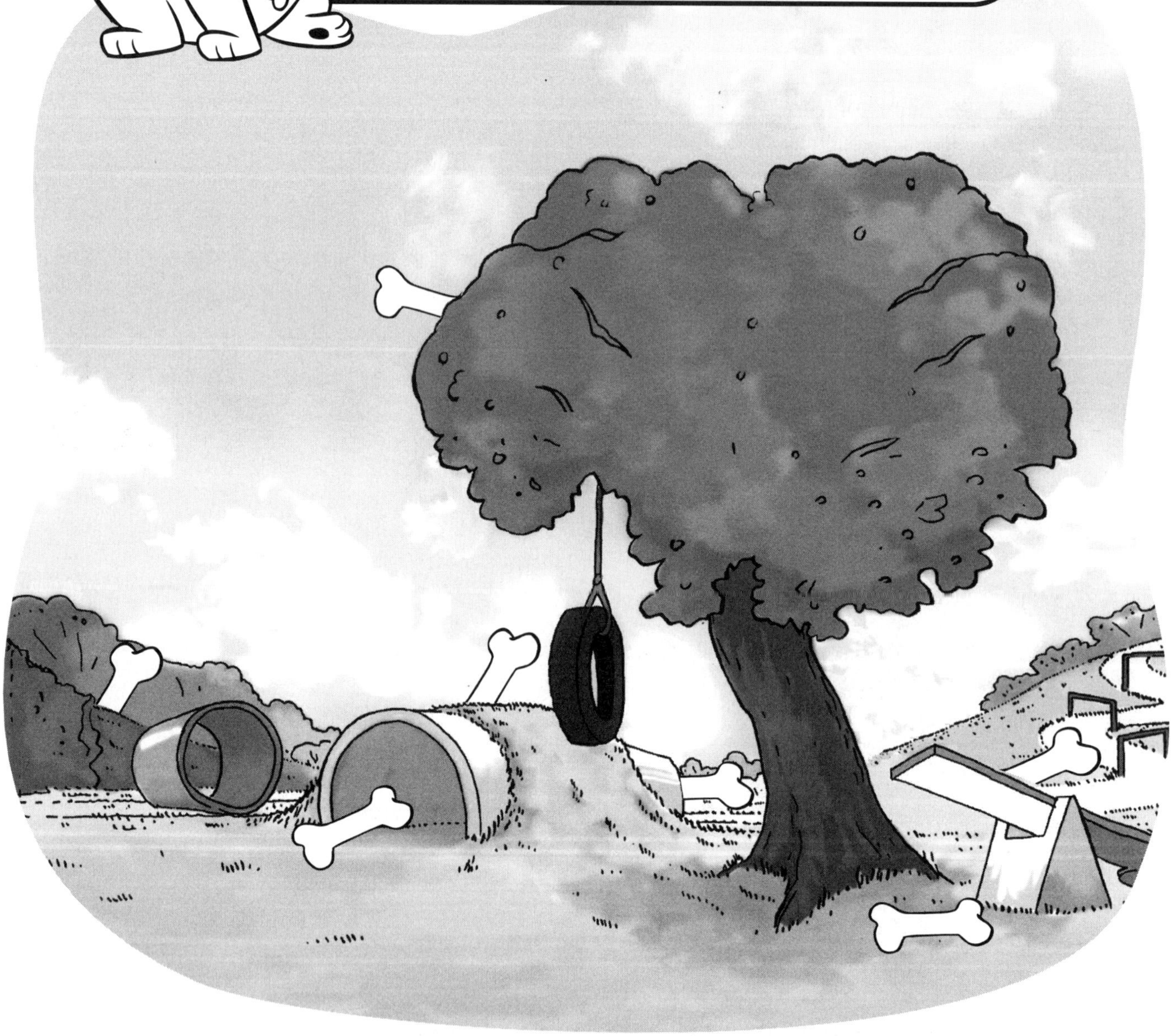

How many bones can you see on this page?

Write the number.

1 2 3 4 5 6 **7** 8 9 10 11 12 13 14 15 16 17 18 19 20

■SCHOLASTIC
PHOTOCOPIABLE

How many letters?

Clifford has lots of letters in his name.
Colour all the letters.

Clifford

How many letters are in Clifford's name?

Write the number.

How many letters are in your name?

Write the number.

1 2 3 4 5 6 7 **8** 9 10 11 12 13 14 15 16 17 18 19 20

Count the shapes

Let's count shapes together.
Colour all the circles red.
Colour the squares blue.
Colour the triangles yellow.

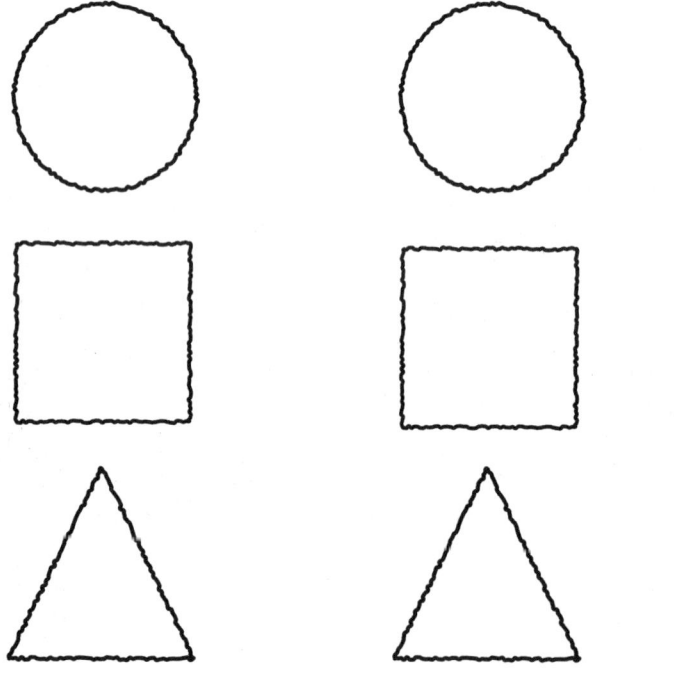

How many circles can you see on this page?

How many squares?   How many triangles?

Count all the shapes on this page.
How many shapes can you see?

Write the number.

Chasing butterflies

Cleo is chasing butterflies. Can you find 10 butterflies on this page?
Circle the butterflies.

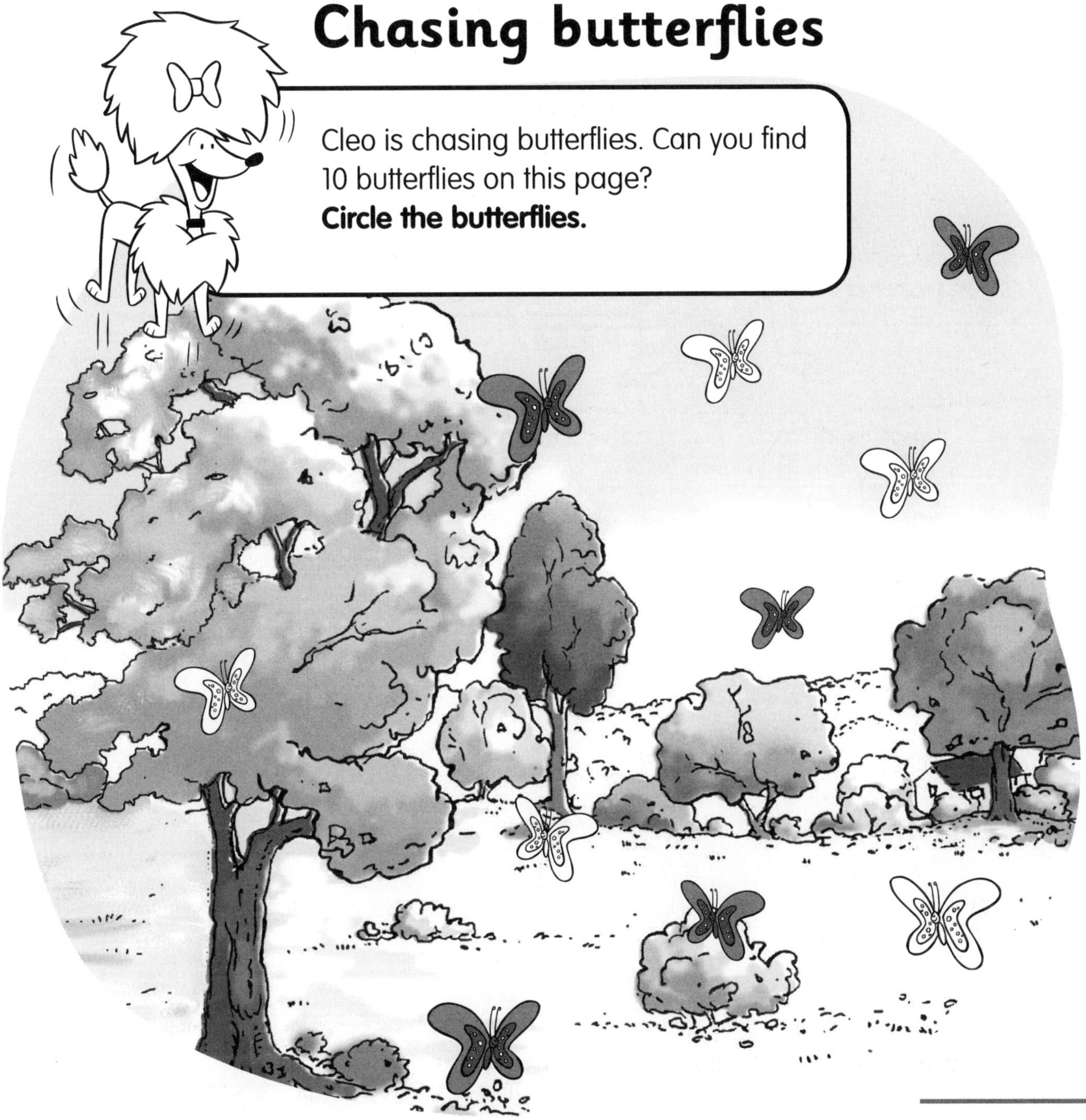

How many **dark** butterflies can you see? **Write the number.**

How many **other** butterflies can you see? **Write the number.**

Count all the butterflies on this page.

Write the number.

1 2 3 4 5 6 7 8 9 **10** 11 12 13 14 15 16 17 18 19 20

Review 6–10

Help me to count the balloons in each group.

Write the number under each group of balloons.

_____
- - - - - - - - - -
_____

_____
- - - - - - - - - -
_____

_____
- - - - - - - - - -
_____

_____
- - - - - - - - - -
_____

_____
- - - - - - - - - -
_____

1 2 3 4 5 6 7 8 9 10 11 12 13 14 15 16 17 18 19 20

■SCHOLASTIC
PHOTOCOPIABLE

Follow the path

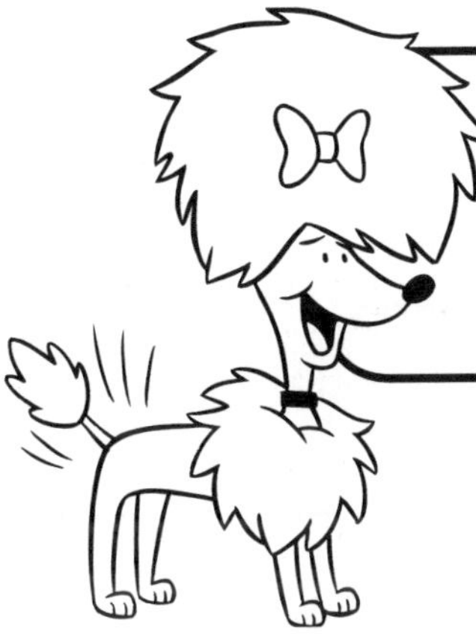

Cleo is looking for T-Bone. She needs to follow the path with the most stones.

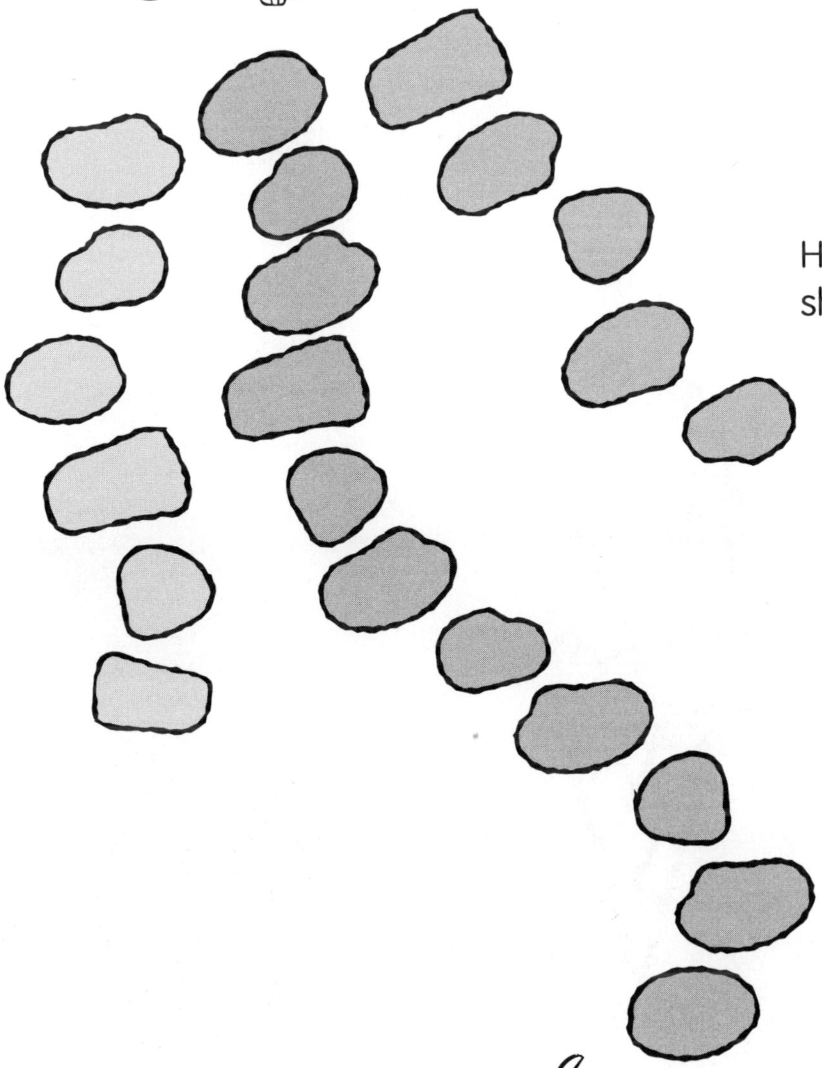

How many stones are in the shortest path?

_____
- - - - - - - - - - - - - -
_____

Now count all the stones in the longest path.
Write the number.

_____
- - - - - - - - - - - - -
_____

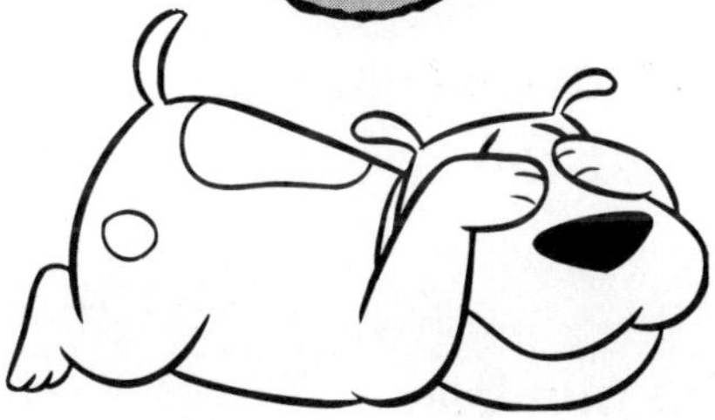

1 2 3 4 5 6 7 8 9 10 **11** 12 13 14 15 16 17 18 19 20

Super scoops!

Look at these ice-cream cones!
How many scoops of ice-cream are on each cone?
Draw one more scoop of ice-cream on each cone.
Now how many scoops of ice-cream are on each cone?

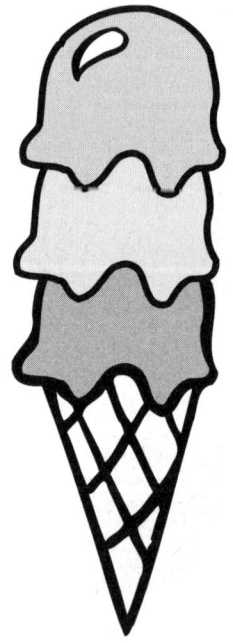

Now let's count all the scoops of ice-cream that are on this page. (Don't forget to count the scoops you have drawn!)

Write the number. _____

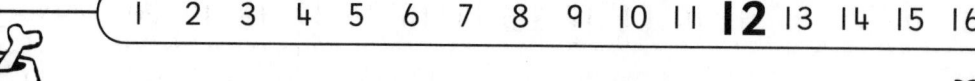

1   2   3   4   5   6   7   8   9   10   11   **12**   13   14   15   16   17   18   19   20

Presents galore!

T-Bone and Cleo have found something fun!
**Draw lines between the numbers to see what it is.
Then colour it in.**

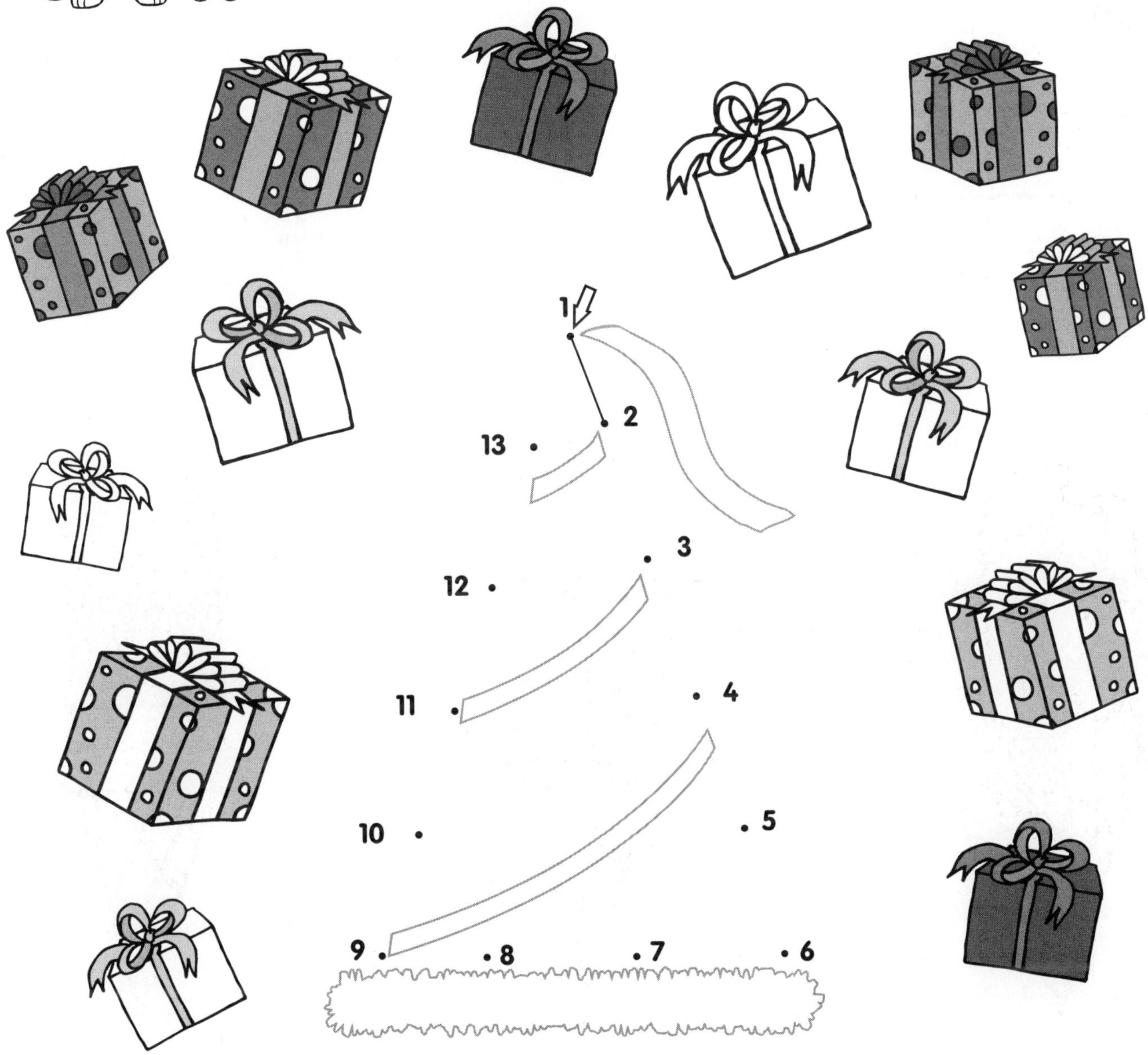

1

2

13 ·

3

12 ·

11 ·   · 4

10 ·   · 5

9 ·   · 8   · 7   · 6

Count the birthday presents.

_____

Write the number.   - - - - - - - - - - - - - - -

_____

| 1 | 2 | 3 | 4 | 5 | 6 | 7 | 8 | 9 | 10 | 11 | 12 | **13** | 14 | 15 | 16 | 17 | 18 | 19 | 20 |

Juggling fun

Clifford and I are trying to juggle balls.

Colour the balls red.
Write the number.

$+$

Colour the balls blue.
Write the number.

$+$

Colour the balls green.
Write the number.

$=$

Count all the balls on this page.
Write the number.

1 2 3 4 5 6 7 8 9 10 11 12 13 **14** 15 16 17 18 19 20

■ SCHOLASTIC
PHOTOCOPIABLE

Match the puzzles

I'm sorting puzzle pieces.
**Draw a line from each set of pieces to
the puzzle that matches.**

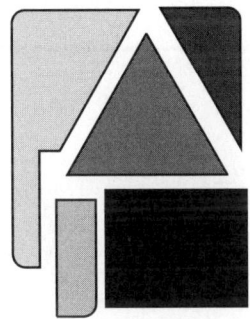

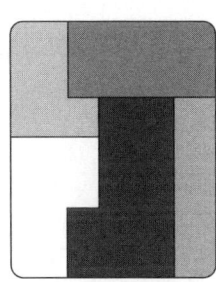

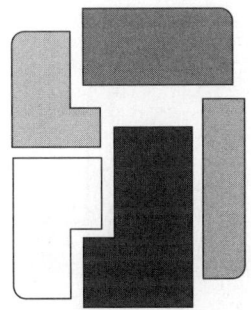

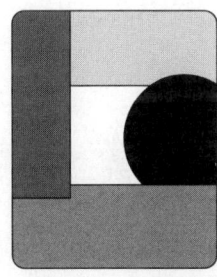

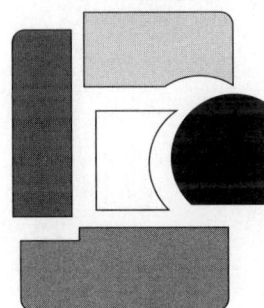

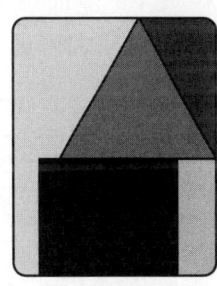

How many puzzle pieces are in each puzzle?
Write the number.

Count the pieces in all three finished puzzles.
Write the number.

1 2 3 4 5 6 7 8 9 10 11 12 13 14 **15** 16 17 18 19 20

Name: Date:

Review 11–15

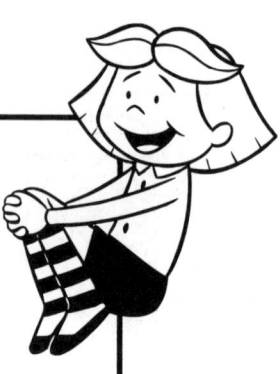

Can you help me to count the stars on my new blanket? How many stars are in each row?

Write the number at the end of each row.

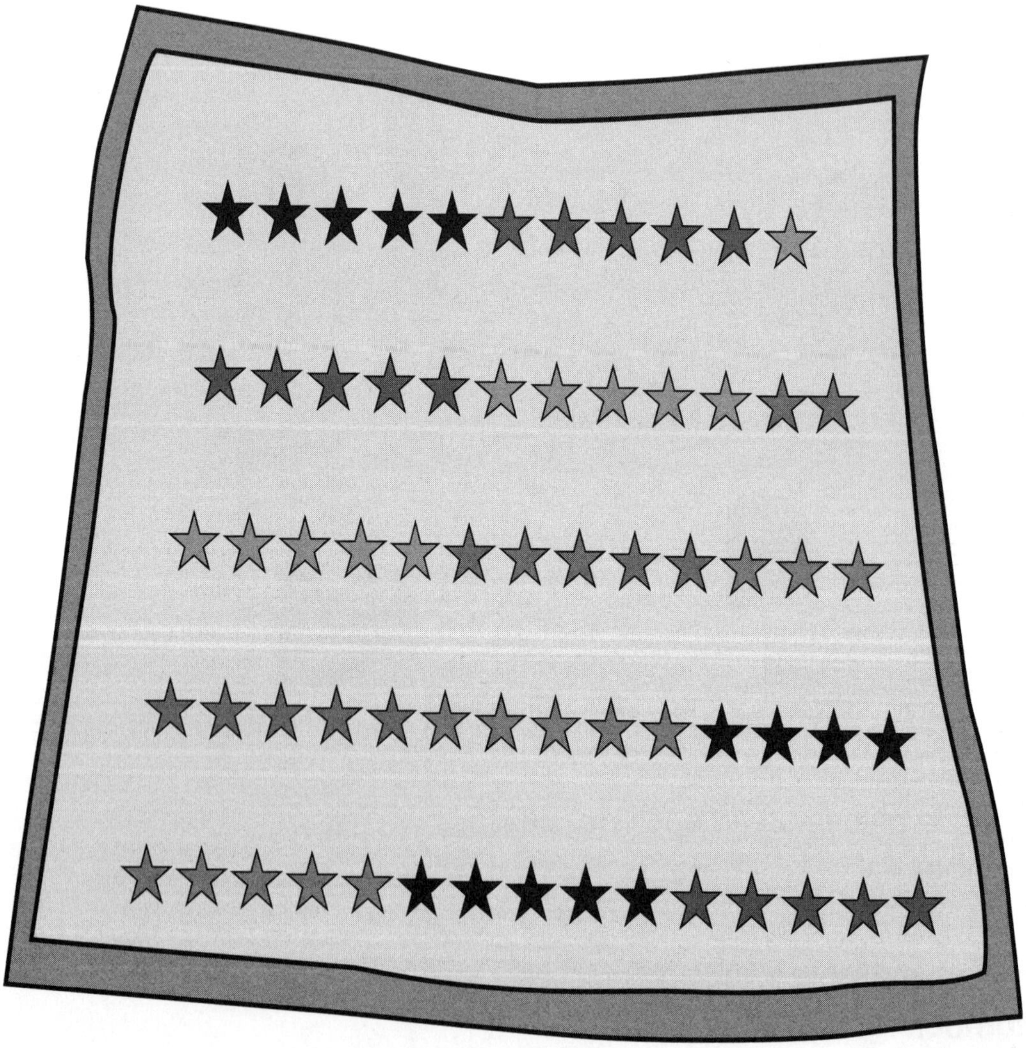

1 2 3 4 5 6 7 8 9 10 11 12 13 14 15 16 17 18 19 20

SCHOLASTIC PHOTOCOPIABLE

Up in the clouds

T-Bone is dreaming that he can fly around these numbers.

Follow the numbers from 1 to 16.
What picture did you make?

Count all the clouds on this page. Don't forget to count the one you made.

Write the number. _ _ _ _ _ _ _ _ _ _ _ _ _ _ _ _

| 1 | 2 | 3 | 4 | 5 | 6 | 7 | 8 | 9 | 10 | 11 | 12 | 13 | 14 | 15 | **16** | 17 | 18 | 19 | 20 |

Missing numbers

Help me to write the number that
is missing in each row.

14, 15, 16, 17

15, 16, _____ ,18

16, _____ , 18, 19

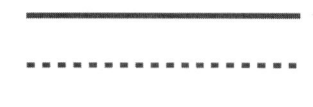

_____ , 18, 19, 20

Count all the hearts on this page. **Write the number.**

_____

1 2 3 4 5 6 7 8 9 10 11 12 13 14 15 16 **17** 18 19 20

SCHOLASTIC
PHOTOCOPIABLE

Match the shells

Clifford is looking for shells on the beach.
Draw lines between the shells that are the same.

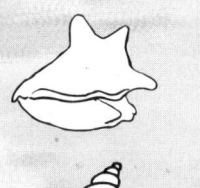

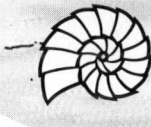

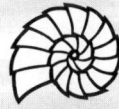

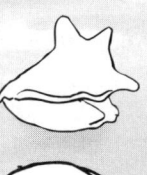

Count all the shells on this page.

Write the number.   - - - - - - - - - - - - - - - -

1 2 3 4 5 6 7 8 9 10 11 12 13 14 15 16 17 **18** 19 20

Starry sky

Clifford's friends like to look at stars.
Nine of these stars look just the same.
Draw circles around them.

Count all the stars that do not have
circles around them.
Write the number.

Now count all the stars in the sky.
Write the number.

1 2 3 4 5 6 7 8 9 10 11 12 13 14 15 16 17 18 **19** 20

◣ S C H O L A S T I C
PHOTOCOPIABLE

Falling leaves

Look at all those leaves!
**Let's follow the numbers from
1 to 20 to see where they are coming from.**

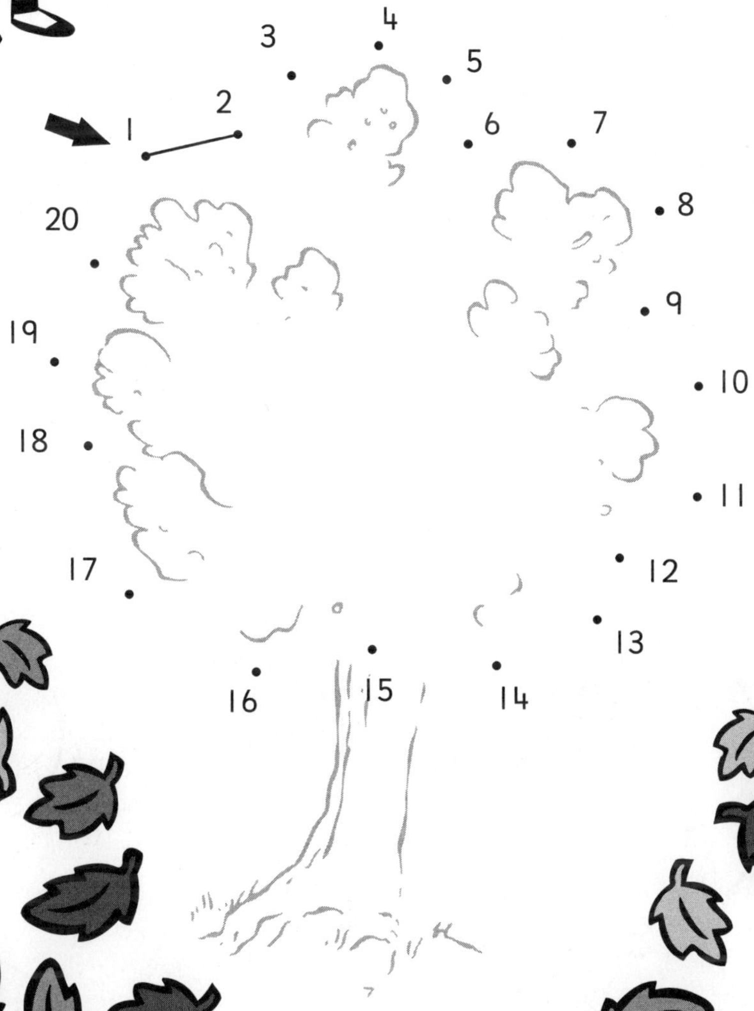

3   4
5
1   2   6   7
8
20   9
19
18   10
11
17   12
13
16   15   14

How many leaves are in this pile?   How many leaves are in this pile?

Write the number.   **Write the number.**

How many leaves are there altogether?

Write the number.

1 2 3 4 5 6 7 8 9 10 11 12 13 14 15 16 17 18 19 **20**

Complete the puzzle

Clifford and I are doing a puzzle.
You can help us. **Draw lines from
the missing pieces to the places
where they belong.**

1	2	3	○	5
6	7	○	9	10
11	○	13	14	15
○	17	18	19	20

8

16

12

4

1 2 3 4 5 6 7 8 9 10 11 12 13 14 15 16 17 18 19 20

◼ SCHOLASTIC
PHOTOCOPIABLE

Count the bones

T-Bone is putting bones into piles.
Count the bones in each pile.

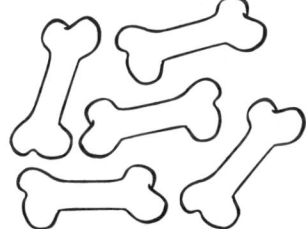

How many bones? _____

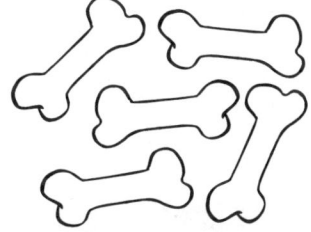

How many bones? _____

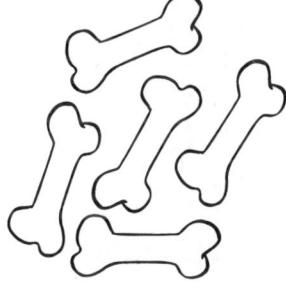

How many bones? _____

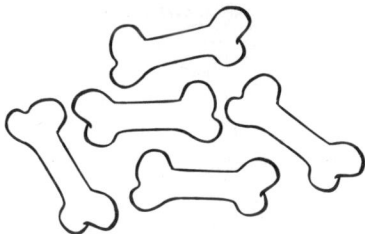

How many bones? _____

Now count all the bones on the page. **Write how many.** _____

1 2 3 4 5 6 7 8 9 10 11 12 13 14 15 16 17 18 19 20

Writing numbers (1)

Let's write number words.
Colour each number. Then trace the number word below it.

1
one

2
two

3
three

4
four

5
five

6
six

7
seven

8
eight

9
nine

10
ten

11
eleven

1 2 3 4 5 6 7 8 9 10 11 12 13 14 15 16 17 18 19 20

◥ S C H O L A S T I C
PHOTOCOPIABLE

Writing numbers (2)

12
twelve

13
thirteen

14
fourteen

15
fifteen

16
sixteen

17
seventeen

18
eighteen

19
nineteen

20
twenty

1 2 3 4 5 6 7 8 9 10 11 12 13 14 15 16 17 18 19 20

SCHOLASTIC PHOTOCOPIABLE

Drop a penny

Clifford and I play games all the time.
Here is a game that you can play.
**Drop a penny on this page. What number
did it land on? Hold up that many fingers.**

1

2   3

4   5   6

7   8   9   10

1   2   3   4   5   6   7   8   9   10   11   12   13   14   15   16   17   18   19   20

📖 SCHOLASTIC
PHOTOCOPIABLE